Industrial Shift

DOI: 10.1057/9781137340313

Also by Joe Atikian

CUBA UNDER EMBARGO: The Macro Impact
SAVING MONEY: The Missing Link

DOI: 10.1057/9781137340313

palgrave▸pivot

Industrial Shift: The Structure of the New World Economy

Joe Atikian

palgrave macmillan

DOI: 10.1057/9781137340313

INDUSTRIAL SHIFT

First published in 2013 by
PALGRAVE MACMILLAN®
in the United States—a division of St. Martin's Press LLC,
175 Fifth Avenue, New York, NY 10010.

Where this book is distributed in the UK, Europe and the rest of the world, this is by Palgrave Macmillan, a division of Macmillan Publishers Limited, registered in England, company number 785998, of Houndmills, Basingstoke, Hampshire RG21 6XS.

Palgrave Macmillan is the global academic imprint of the above companies and has companies and representatives throughout the world.

ISBN: 978-1-137-34027-6 EPUB
ISBN: 978-1-137-34031-3 PDF
ISBN: 978-1-137-34226-3 Hardback

Library of Congress Cataloging-in-Publication Data is available from the Library of Congress.

A catalogue record of the book is available from the British Library.

First edition: 2013

www.palgrave.com/pivot

DOI: 10.1057/9781137340313

Contents

DOI: 10.1057/9781137340313

DOI: 10.1057/9781137340313

List of Tables

DOI: 10.1057/9781137340313

List of Figures

DOI: 10.1057/9781137340313

DOI: 10.1057/9781137340313

DOI: 10.1057/9781137340313

1
Introduction

Abstract: *Are food and goods "no longer made here"? Increased trade makes it seem so: India 1960s, Japan 1970s, China 1980s, Russia 1990s. Globalization and employment shifts bring fear of Western decline. The shift from farms to factories reinforces the fear, although economic data showing their growth seem to be ignored. The service sector grows much quicker and gives the illusion of decline in the other sectors.*

The most advanced economies have the smallest share of GDP in agriculture, even while they grow ever more food. They also have the highest incomes.

Countries with less advanced technologies grow by shifting to manufacturing. Wealth grows slowly at first, but increases rapidly as they shift further from agriculture. The same pattern takes place in most countries around the world.

Key words: industrial shift; industrial structure; manufacturing sector; productivity; service sector

Atikian, Joe. *Industrial Shift: The Structure of the New World Economy*, New York: Palgrave Macmillan, 2013. DOI: 10.1057/9781137340313.

The entire world economy seems to be undergoing another rapid, fundamental shift. A recent series of monumental changes within some of the world's largest nations has heightened the attention on the structure of their economies. India's green revolution of the 1960s turned the country from a dependence on food imports to a net exporter. In 1980 China regained America's Most Favored Nation (MFN) status in trade, and has since vaulted toward superpower status. The Soviet Union collapsed in 1990 and turned its closed trading system outward. Japan's economy, though nearly stagnant since 1991, has greatly increased foreign investment in a bid to offset the effect of a highly valued yen. Finally, Europe saw a reunified Germany, and built a major currency union.

A defining feature in this cluster of events is an increase in global trade enabled by liberalized treaty regimes—the GATT and WTO. Meanwhile, this trend toward increasing global trade has added to existing worries within some of the major Western economies that they are losing capacity and skills in agriculture, and, especially, in manufacturing. One increasingly popular viewpoint has it that 'food and goods are no longer made here'. But, is it true?

GDP is certainly growing in most countries, so if food and goods are made elsewhere, then where does growth come from? Only three sectors exist: agriculture, manufacturing, and services, so if manufacturing is in decline, then which sector drives GDP growth? Agriculture never springs to mind as a growth driver, so by elimination, growth must be driven by services. But the services sector supposedly offers low value, low pay, and low productivity growth. So where does growth come from? Contrary to popular perception, neither agriculture nor manufacturing is in decline.

The widely publicized loss of family farms is real, and it reinforces the anxiety about feeding a growing world population. Meanwhile the ongoing loss of manufacturing firms gets extensive media coverage, and creates political pressure to protect affected industries from declining output and employment. But the loss of farms and factories turns out to be only half of the story. Small farms are indeed declining in Western countries, but their operations merge and continue producing. Textile manufacturers close or move overseas, but in their place, makers of aircraft and medical instruments arise. Jobs are lost but they are also gained; the broader issue is how these add up. China stands virtually alone as the emblem of shifting industrial structure, and has become a prime target of American angst about job losses. But, contrary to

DOI: 10.1057/9781137340313

popular opinion, the general trend in US unemployment has improved as trade with China grew, even with a rising trade deficit. Some of the signs are contradictory, so rather than relying on a loose impression of these important industrial shifts, a fuller view of the economy's overall condition demands an understanding of both losses and gains.

As the global economy evolves, each country shifts its focus between farms, factories, and offices. These shifts in the composition of GDP have been widely recognized since at least the 19th century. Everyone sensed that manufacturing played an increasingly important role compared to farming, but change was slow and international data were scarce. Policy decisions relied on crude estimates or strictly political judgment. The past 40 years have brought a revolution in data reporting that not only confirms the suspected trends, but reveals several surprising developments. Most countries around the world have continued to decrease their economic concentration on agriculture. Meanwhile the other two sectors and overall economic output continued rising. Today, only the Least Developed Countries still rely heavily on agriculture. Conversely, the most highly developed countries have nearly eliminated agriculture as a significant contributor to GDP. But most surprisingly, manufacturing is no longer the best indicator of growing national wealth.

Further discoveries arise in the patterns underlying this structural shift. Only the most industrially advanced countries have, for the first time in history, stabilized the share of agriculture in their economies at a very low level. For these advanced countries, the relative size of their farming sector tends to fall gradually, and in many cases stops at a mere 2%. Among countries nearer the beginning of their industrial transition, those that shift very quickly away from agriculture achieve the highest GDP per capita growth rates. Of course, this implies that countries with a slower shift in industrial structure are prone to suffer slower growth in GDP per capita. In general, the speed of technological progress is a good indicator of advances in wealth and living standards.

A less obvious feature is that the amounts of agriculture and wealth have a contrary relationship. In a less advanced economy, one that relies heavily on agriculture, wealth at first grows slowly with industrial progress. So, large shifts away from agriculture come with small increases in wealth. Farming may seem like a trap, as the gains from shifting away are small and risky. As the nation's industry advances though, wealth rises more rapidly with each step away from agriculture. Finally, the

DOI: 10.1057/9781137340313

agricultural share settles to a low and stable level while wealth continues to rise. At this point, manufacturing may stabilize its share of GDP, or even begin to decline. In this most advanced state, the real output (not the share) of the three major industrial sectors keeps growing and improving.

Notwithstanding the general decline in the agricultural share of GDP, this sector is neither technologically stagnant nor shrinking in absolute output. Productivity remains high and continues to improve. Even as the agricultural share steadily declines, this sector's total real output continues to grow worldwide, outpacing population growth by a wide margin.

The same transition path applies to economies of all sizes, and closely resembles the path of entire regions, and even the whole global economy. Some aspects of these broadly based patterns suggest that poor but progressing countries are on a similar economic path to that previously experienced by wealthier countries. The recent shift in manufacturing capacity to Asia shows this most clearly. So although the slow rise in wealth can be disappointing in the early phases of development, an increasingly rapid rise seems likely.

In the 1940s economists thought that there were limits to concentration within the three main industrial sectors. It seemed clear that there should be a limit to the share of services in GDP, at least because services were difficult to trade, but also because a country was thought to depend crucially on growing its own food and manufacturing its own tangible products. The ensuing decades of data showed that this was not to be the case. It turned out that services could be as easily traded as hard goods. For example, engineering services can be conducted by overseas firms, and the resulting calculations communicated electronically. Tourism, banking, technical support, and many other services are now routinely traded, facilitated by computers, the internet, and low-cost air travel. These might have been difficult to imagine in the 1940s, and their existence today shows that prediction should be approached only with great caution.

One of the clearest trends over the past four decades is the settling of agricultural concentration to a plateau of about 2% of GDP in the advanced economies. The data are consistent across many countries, tempting observers to declare a natural limit, below which the share of agriculture will not fall. In another clear trend, global manufacturing has remained at a stable share of GDP over the same period, although rising in some regions and falling in others. But just as it was once impossible

DOI: 10.1057/9781137340313

to foresee the international trading of services, it is now difficult to imagine a successful economy with a declining share of manufacturing. Even though such a change might be unsettling, there is no necessity that an advanced country's industrial structure keeps shifting along a steady path of rising manufacturing share. That's because for most countries at most times, their economy is not a closed system. There is usually interaction with other countries through trade. That means that one country's structure can shift in reaction to changes in other countries. So for example, as China's manufacturing sector grows, its enormous size will affect other economies. Others can transfer manufacturing to China, and focus increasingly on services. And this can go on for a long time, even to the point where there is near total specialization, such as China focusing on manufacturing, and the US focusing on services.

Eventually though, the logic of industrial shift will see those newly transformed countries change yet again. The process of specialization within a more mature economy followed a path such that it shifted focus from agriculture to manufacturing to services. That was because people do not sit idly and replicate agricultural systems without improvement; nor do they manufacture products in the same way generation after generation. Instead they seek product improvements and process efficiencies, which feed the shifts into other sectors. Likewise, a manufacturing focused China will seek to improve its processes and products, which involves specialization and transferring many activities to its own services sector. This can include transferring several traditional activities out of manufacturing firms and into services firms that specialize in areas such as administration, maintenance, transportation, or warehousing.

As these changes proceed, a backlash can develop against the trend. Currency values may rise in manufacturing economies, driving production back to traditional locations. Transportation costs may rise with oil prices, having a similar effect. In past episodes of rising globalization, the trend had reversed itself due to political pressure. Public opinion often turned against expanded trade, resulting in import restrictions and other actions that were seen to encourage domestic industries. This line of thinking is still held today, and it may grow sufficiently popular to suppress the West's shift away from manufacturing. Alternatively, it may turn out that China's economy grows far larger than America's in the coming decades, providing an outlet for ever rising American services exports. So instead of the grim vision of a declining Western world, a resurgent East can be a new growth opportunity. Over the past twenty

DOI: 10.1057/9781137340313

years, about 30% of US export activity was made up of services, including R&D testing, engineering, architecture, insurance, finance, and business management.[1] There seems to be no reason that services cannot become a majority of a country's export economy, especially to the burgeoning population centers such as China, India, and Brazil.

The great industrial shift since about 1900 was the drastic reduction in agriculture as a share of the world's total output. With new technologies and rising productivity, industry changed its basic character and composition as farms gave way to factories. Even though manufacturing output continues to grow, the new shift is that manufacturing is no longer a rising share of most of the advanced economies; nor is it a rising share of the entire global economy. This constitutes a major step toward the new world economy.

Note

1 U.S. Department of Commerce, Bureau of Economic Analysis, International Economic Accounts, *Trade in Goods and Services 1992–present.* http://www.bea.gov/international/index.htm#trade. Also detailed statistics for cross-border trade: *Trade in services 1999–2011.* http://www.bea.gov/international/international_services.htm. Accessed January 5, 2013.

DOI: 10.1057/9781137340313

2
What Is Industrial Structure?

Abstract: *The term "industrial structure" can be used in a few ways. This chapter lays out a straightforward description of the term as a classification of the three main economic activities: the agricultural, manufacturing, and services sectors. A pivotal concept is defined: the share of each sector in GDP. As the services sector grows, the other sectors appear to shrink. This shift is, of course, merely in the relative size of each sector, and not necessarily a decline in real output.*

Through this discussion, it becomes apparent that the supposed decline in manufacturing is partly a mere artifact of the data reporting system. This helps to plant the idea that the popular image of decline does not accurately reflect the working economy.

Key words: industrial

Atikian, Joe. *Industrial Shift: The Structure of the New World Economy*, New York: Palgrave Macmillan, 2013. DOI: 10.1057/9781137340313.

Extract, transform, deliver

Industry produces many of the goods and services that people in modern societies want and need. Although people also have non-economic needs such as families, social connections, and freedoms, these are not produced, traded, stockpiled, or priced in a systematic way. So economic products are a distinct category that people concern themselves with, the products of an industrial system.

Industrial structure describes the composition of a country's economic activity, the production of human material provisions. Industries are usually categorized into three basic types according to their stage within the production process, or the type of value being added to a natural resource.

Primary industries are called extractive, as they add value by removing raw materials from the natural setting. This category includes agriculture, fishing, forestry, and hunting; it uses the main biological resources in the natural world. The main resources extracted by these industries are continually cycled through plants and animals, including carbon from the air, all forms of water, and minerals from the soil. Some people view agriculture as a major exporter of water, as well as a consumer of soil, which confirms the extractive aspect of the category.

Some confusion exists about the relation of farming and food in this category. Because farms can supply many other industries as well as consumers, their products are most often tallied before being processed. Apples sold to consumers count as farm produce, but apple pies count as a manufactured product. In a related example, a farm can grow cotton, which is used to produce t-shirts, but only the cotton is counted as a farm product. The end result is that in the US for example, farm output is about 2% of GDP, whereas the 'food and fiber' sector that uses farm products makes up about 16% of GDP.

Secondary industries involve transforming raw materials into usable products. These are the manufacturers, many of which are classified as intermediate industries if they make products for other industries. For instance, steel makers never sell their products directly to consumers, so they fall into this intermediate class. Within the handful of systems used to categorize industries, the secondary category includes mining because, even though mining appears extractive, it shares many other characteristics with manufacturing. Miners do not stop at digging mineral-laden rock from the ground. Instead, they usually process the

DOI: 10.1057/9781137340313

raw material by crushing, separating, and concentrating it into a more refined form to avoid shipping vast amounts of unusable rock.

Tertiary industries include those generally non-physical services that provide intermediate support to firms, and direct services to consumers. These include, for example, accounting, public relations, and marketing for firms, in addition to various consumer services such as education, medical care, and entertainment.

One widely used categorization system is the UN's International Standard Industrial Classification of All Economic Activity (ISIC).[1]

Guns, butter, and haircuts

'Guns and butter' is a phrase often used to illustrate the inevitable trade-offs in economics. People want many kinds of goods, like food and security, but with finite resources and time available, they can only produce a limited amount of each good. So, choosing to produce more butter implies that they will have fewer guns. The total amount of goods and services produced by a country's industries each year is called its gross domestic product (GDP). But, although GDP is useful as a measure of the economy's overall size, it cannot indicate the kinds of goods and services produced. In other words, GDP is an aggregate so it cannot show the share of guns and butter produced within the economy. Turning to the idea of Industrial Structure, guns and butter can be seen in another useful way. They can be classified as manufactured goods and agricultural goods, the first two economic sectors. Economists sometimes introduce the third and final sector of the economy by referring to haircuts. Haircuts are a service because there is no tangible product involved; they cannot be imported or exported as the other goods can be. The output of every economy is composed of these three basic sectors.

While GDP shows the size of a national economy, its growth can be affected by luck and by deliberate manipulation. Russia and Canada chanced upon extra GDP growth in the 1990s, caused in part by rising commodity prices. Meanwhile, several Latin American countries tried to force growth using tactics such as expanding money supply, import–export regulations, currency manipulation, and capital controls. While the effects differ in each case, they were usually short lived because the underlying structure of the economy remained unchanged. Short-term

DOI: 10.1057/9781137340313

growth can sometimes be imposed by such policy tools, but stable growth rests on another kind of foundation.

Although GDP growth is an important indicator of national wealth, it does not show just what kind of progress is being made within the economy. Did the economy grow merely by shipping more of the same natural resources, or did it grow due to the advancing skills and capabilities in its industries? Did it grow merely due to rising prices for the same crops year after year, or did some farmers shift to higher value manufacturing? Did it grow merely by making more of the same furniture and textiles, or by shifting to satellites, medical imaging, and engineering services? In short, growth in the economy could be driven by expanding its traditional industries, or by doing new things in better ways. The former path carries the risk of being overtaken by other innovative countries, and the risk of declining living standards. So the degree of development, ranging from industrial stagnation to continuous industrial shift, carries major implications for the well being of every country's population.

Canada and the United States struggle with these questions daily. The Canadian prime minister advocates building new pipelines to export crude oil to Asia, while opponents demand that it be refined at home in order to maximize the economic benefit. The US president advocates rebuilding the manufacturing sector using new trade legislation, while free-traders warn of foreign retaliation and higher prices if he takes that path. For both Canada and the US, the reality is that resources and manufacturing no longer play the most important role in the economy. Both sectors still matter, and both are still growing in size, but the services sector now dwarfs them. Knowing where the economy's strength lies begins with knowing its structure. And understanding the world's structure reveals where the opportunities might lie.

A graph of GDP can show only the economy's growth rate, whereas a glance at a simple graph of its structure reveals both the pace and the quality of its progress. Periods of political difficulty as well as economic recession and technological stagnation stand out clearly. Industrial structure plainly reveals a country's economic progress, and resists artificial maneuvering by governments or statistical agencies. Examples in Chapter 4 compare the erratic industrial shift in Argentina with the smooth shift of South Korea over the recent decades. They show a concise way to measure the condition of an economy, and demonstrate why industrial structure should become a standard measure of the economic status of nations.

DOI: 10.1057/9781137340313

Distortions and perceptions

Nobel laureate economist Simon Kuznets identified a few key problems with analyzing industrial structure early in the 20th century. He saw the problems as serious enough that some of the world's most important countries could not be properly compared. Although China was barely beyond the agricultural era, it was considered highly interesting owing to its great size and formidable history. After all, the opening of Japan to world trade in the 1850s unleashed a powerhouse from a relatively small nation within a few decades. One might only have imagined the impact of an industrialized China. But the US and China were at such different stages of industrial advancement that their economies could not even be conceived in the same terms. So the practical problems in this field are well known. For example, Kuznets explains that many services that are routinely supplied by firms within developed economies are normally unpaid family activities within the less developed economies. He mentions goods such as manufactured food, clothing, and furniture, and services such as insurance and funerals, which form part of the American industrial base, but remained as household or community duties in China. (These so-called non-economic products mentioned above can obviously develop into economic ones.) These activities defy easy classification, resulting in an overstatement of the output gap between nations.[2] In turn, this weak information limited the ability to formulate trade, financial, and other policies.

Much of Kuznets's worry has been solved by two major developments. First, most of the major economies in the world have progressed to the point that much of their industrial activities are of similar types, and can be compared. Second, with all but a small handful of countries, consistent economic data are coordinated and reported to global agencies, primarily the United Nations. Similar institutions include the IMF and World Bank, along with more focused agencies such as the OECD, which share much of the UN's data.

But there is also another type of problem that goes largely unrecognized. Specialization within firms skews the economy's structure away from the manufacturing sector and toward the services sector by outsourcing various activities. Activities such as business management, accounting, marketing, payroll, and numerous others are being continuously diminished within manufacturing firms. As specialized firms arise

DOI: 10.1057/9781137340313

and perform these functions, their output counts toward the expanding services sector even though the entire benefit of those specialized services still goes to the manufacturing firm.[3] In effect, manufacturing merely *seems* to be in greater decline and services merely seem to be in greater expansion.

Of course the same specialization and outsourcing process occurs in the agricultural sector, but that interaction involves manufacturing in addition to services. This results in a cascade of interaction. Farms can outsource some administrative functions, such as accounting, which increases the service sector output and decreases the relative size of the farm sector. Farms can also, for example, augment labor with purchases of manufactured goods such as fertilizers, trucks, and harvesters. To the extent that these goods reduce costs, they can also help to reduce farm selling prices and thus the size of the entire sector. The purchase of manufactured goods further increases the manufacturing sector's output, which further decreases the relative size of the farm sector. In this scenario, farms produce more and more goods, while the agricultural sector registers an ever decreasing share of GDP.

The declining share of agriculture may end up being more exaggerated than that of manufacturing due to its greater extent of interaction. But, even if the reported declining share of agriculture is greatly exaggerated, there is little practical effect. The sector still increases its specialization, productivity, and output.

Naturally, since the decline in agriculture and manufacturing is exaggerated, the rate of growth in the services sector must also be overstated. There is no doubt that entirely new types of services have contributed to service sector growth. These are services that were not previously done within manufacturing firms, such as web hosting, leisure travel, family law, cosmetic surgery, film production, or software development. But it is also certain that many services have merely been reclassified away from the older sectors; as such they should not strictly count as growth in services. For instance, outsourced accounting for a manufacturing firm could just as rightly be considered to be a manufacturing sector activity as it could a service sector activity.

All three sectors are deeply intertwined, so it seems clear that outsourcing by firms causes some distortion in the share of each sector. Part of the common perception that manufacturing is in decline comes from this distortion.

Data sources

Several information sources can be used to compare industrial structure, such as the statistical services of the United Nations, World Bank, and OECD. UN statistics are available in a unified format for most countries of the world starting from 1970 and are used here simply because they cover more countries than any other source. Most other reporting agencies are membership organizations with more restrictive inclusion criteria.[4] Although most data are available as late as 2011, a cut-off of 2008 was widely used here to avoid any distortions caused by the recent global recession.

Although there are currently more than 200 countries, most of the following discussion refers to a global total of only 177. This is done in order to exclude states that were formed after 1970, which is the start of the study period. For example, the 1989 disintegration of the USSR and Yugoslavia resulted in several new states that obviously have no historic economic data. Because these states cannot be fairly compared over the same time frame, they are omitted.

As with most reporting agencies, the UN data do not uniformly present agriculture as a discrete component of GDP. Instead, it is combined with forestry, fishing and hunting in a pooled extraction sector. This leads only to a very few exceptional cases, such as Iceland, which shows a very high share of agriculture in its GDP even though it has very little terrestrial farming but a great deal of fishing. Throughout the following chapters, the terms agriculture and farming refer to the entire primary industrial sector.

A few other sources are used wherever more detailed information is needed. For example, the US Bureau of Economic Analysis (BEA) provides historical data on many industries. Likewise the OECD compiles data for a select group of advanced economies, while several other regional organizations publish annual reports. A few countries present some difficulties as they independently revise their statistical practices. For instance, Cuba has occasionally joined and withdrawn from international reporting agencies, while the precision of data from China and North Korea is often contested. The usual response from global agencies such as the United Nations is to publish estimates during periods in which significant doubts arise. Even so, the extent of disagreement on the data tends to be minor, and the overall economic patterns are therefore still thought to be reliable.

DOI: 10.1057/9781137340313

Outline

Industrial structure is usually expressed as the percentages or shares of the three major sectors in a country's gross domestic product. For example, a country's GDP could be composed of 5% agriculture, 25% manufacturing, and 70% services. It could, however, also be expressed more simply by the share of agriculture alone. Because agriculture is the most ancient major industrial sector, the degree of industrial development within a country can be gauged by how far its economy is biased away from agriculture. The agricultural sector forms a key element of the following discussion because it is the most common element among national economies.

Although GDP is a crucial measure of a country's size, it bears little relation to the economy's industrial structure. For instance, many huge economies have a low share in agriculture, but so do many small economies. Instead, the strongest relationship is between structure and wealth; wealth is generally represented as GDP per capita (GDPPC).

In order to outline the industrial structure of the world's economies, three main elements are described:

- the global aggregate of all economies;
- the geographic patterns of industrial structure;
- the relation of industrial structure to wealth (see Figures 3.1, 3.2, and 3.3 respectively).

To illustrate the general idea behind these three points, imagine that:

- agriculture is about 5% of the global economy;
- the regions that have the most agriculture are in the tropics;
- more agriculture is associated with less income per person.

A snapshot of the world's current structure is presented first; this is the world's economy at one point in time. How important is each sector today? Does any single sector dominate? Even though this is merely a snapshot, it provides a starting point from which to visualize the changing world.

This static view is followed by an outline of the shifts in structure and GDP per capita. Are farms and factories in decline? Where are these changes taking place? How fast are the changes happening? These shifts develop at three levels: global, regional, and national. Some important patterns quickly emerge, including links to climate, technology, and historical ties. Several charts and sidebars provide evidence and extended information, although these can be disregarded according to the reader's preference.

Finally, as the growth of technology shifts industries and wealth around the world, it also affects employment. If technology has been growing for hundreds of years, why is everyone not unemployed? Doesn't technology kill jobs? Does trade with China shift jobs from West to East? Did the same kinds of problems exist when agriculture was dominant?

Notes

1 http://unstats.un.org/unsd/cr/registry/regcst.asp?Cl=27&Lg=1
2 S. Kuznets, "National Income & Industrial Structure," *Econometrica*, Vol. 17 July 1949; pp. 208–9; The Econometric Society. (See also S. Kuznets, "Quantitative Aspects of the Economic Growth of Nations," *Economic Development and Cultural Change*, Vol. 5, No. 4, July 1957, pp 1–111; University of Chicago Press).
3 See UN Publication, ISIC Rev. 4 (ST/ESA/STAT/SER.M/4/Rev.4) (2008), Section 5.(a)139 Outsourcing; page 29.
4 All UN data references herein are derived from the UN National Accounts Main Aggregates Database (1990 prices), http://unstats.un.org/unsd/snaama/dnlList.asp (unless otherwise noted).

DOI: 10.1057/9781137340313

3
The Global Status Quo

Abstract: *The global economy is divided into the familiar 5% agriculture, 25% manufacturing, and 70% services. The highest agricultural share is in the tropics, with a lower share in wealthier countries. The wealth of nations can be predicted by a low farm sector share.*

Manufacturing and services have reversed their relation to wealth. Major advances in technology have made services tradable, and now a higher share in services characterizes advanced economies.

Key words: industrial structure; manufacturing sector; poverty of nations; services sector; wealth of nations

Atikian, Joe. *Industrial Shift: The Structure of the New World Economy*, New York: Palgrave Macmillan, 2013. DOI: 10.1057/9781137340313.

 DOI: 10.1057/9781137340313

Global structure

Agriculture is by far the lowest ranking sector in the current global economy, comprising barely 5% of aggregate GDP (see Figure 3.1). Manufacturing ranks second at 26%, while the services sector makes up the heavily dominant balance at 69%. The same general pattern holds true in all regions and most countries. For example, agriculture accounts for the lowest share of South America's GDP at 10.1%, and the lowest share of US GDP, at 2%.

It may be surprising that most of the world's advanced economies derive less than 2% of their GDP from agriculture, even though this 2% includes fishing and forestry. At the opposite end of the spectrum are the least industrially advanced nations, typically small and war torn. These are the most agriculturally intensive countries, with about 62% of their GDP in agriculture. On the fringes, there are only a handful of advanced economies with a high concentration in this sector. These include Iceland (6%), with its focus on fishing, and New Zealand (5.3%), with extensive dairy and meat production focused on export markets. Conversely, not even a single nation on the UN's Least Developed Countries list[1] has less than 3% of its GDP in agriculture (see Appendix A). Agriculture is a good general indicator of industrial advancement and wealth. As a caution, note that a low concentration on agriculture does not mean this sector is small or unimportant. In the advanced and wealthy countries, a low agricultural share means that farm output is still growing, but manufacturing and services are growing faster.

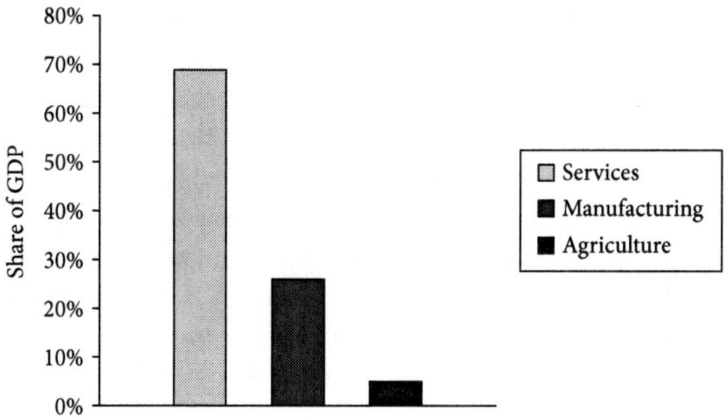

FIGURE 3.1 *Global industrial structure, 2008*

Source: Derived from UN National Accounts Database.

DOI: 10.1057/9781137340313

Geographic distribution

Among all the world's economies, agricultural concentration varies widely. At the low end of the range, Monaco generates absolutely none of its GDP from agriculture. At the high end of the range, Liberia's farm output generates 62% of its GDP (see Figure 3.3). Countries can be fitted into three categories of industrial structure (according to share of agriculture in GDP) that correspond with the stark divisions between the richest and poorest:

- less than 3% (richest countries);
- between 3% and 7% (moderately wealthy countries)
- greater than 7% (poorest countries).

Throughout the world, and, with very few exceptions, more concentration on agriculture means lower overall economic output. Furthermore, a surprisingly narrow margin separates the very rich from the very poor. All of the world's very poor countries derive over 7% of GDP from agriculture, while all of the very rich countries are at less than 3%.

Corresponding to this narrow range in industrial structure is an enormous range of income. The extremes of GDP per capita range from less than $2 per day in heavily farm-dependent Liberia and Laos, to over $180,000 per year in purely non-agricultural Monaco. In the middle ground of industrial structure, between 3% and 7% in agriculture, GDP per capita lies within a moderate range. This GDP gap has created the ongoing economic puzzle of why seemingly slight changes in activity are consistently related to huge changes in output. Five percent in farming seems very similar to 3% in farming, and yet the resulting difference in national income is startling. There are no easy answers, but the first step is to map the pattern using the share of each country's farming sector and its wealth. Figure 3.2 shows this geographic distribution.

Agriculture follows the familiar pattern of North–South wealth distribution. That is, wealthier countries are less reliant on agriculture, and they are clustered outside of the tropics. High income originates almost exclusively in the mid-latitude cooler climates. Conversely, those countries more reliant on agriculture tend to be poorer and grouped within the low latitude hot tropical climates.

The lowest category of agricultural share in GDP, those countries with less than 3%, is located primarily in the wealthiest parts of North America and Western Europe. This group also includes twelve of the

DOI: 10.1057/9781137340313

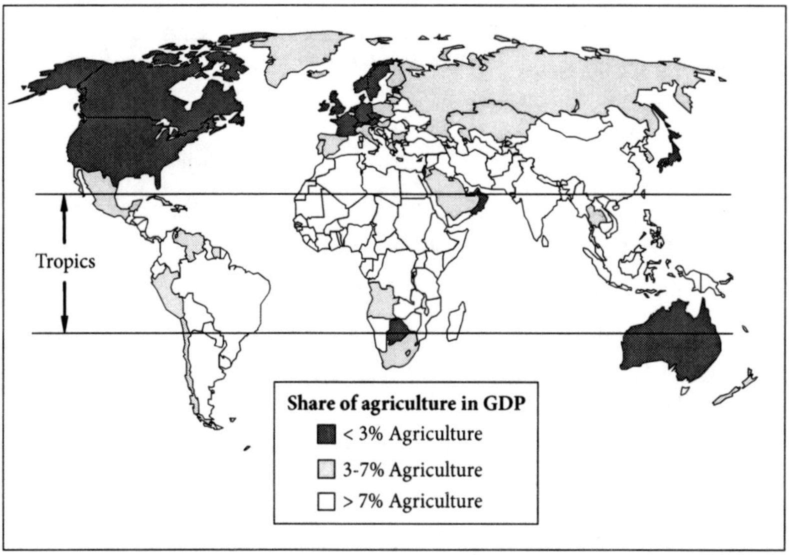

FIGURE 3.2 *Agricultural share in GDP, by country, 2008*
Source: Derived from UN National Accounts Database.

Caribbean islands and six of the major oil exporters. Only one-third of these 44 countries are located in the tropics. All of the wealthier tropical countries are tiny islands, the largest being Hong Kong with a population of about 7 million. Their GDP per capita runs to triple the next category, those middle income states with 3% to 7% of GDP in agriculture. This middle group consists of small to medium-sized countries scattered across all regions of the world, plus Russia. Fully one-half of the middle income countries are in the tropics.

The third and largest group has the highest share in agriculture, with each country at more than 7%. The poorest regions, Africa and South Asia, produce 17% of their GDP from agriculture, more than triple the global average. Fully three quarters of these developing countries are in the tropics. They range from countries of vast land mass such as Sudan, to those miniscule in land mass, population, and total economic output such as Haiti. These include most of the poorest countries in South America, Africa, and Asia, as well as a number of tropical islands, and former Soviet republics. Two notable exceptions are Brazil and Argentina with a rare combination of relatively high GDP per capita, and a high share of GDP in agriculture. The relatively wealthy OECD

DOI: 10.1057/9781137340313

TABLE 3.1 *Agricultural concentration vs. output, 2008*

Share (%)	# of Countries	Average GDP per capita ($)	Region
< 3	44	28,614	N. America, Euro, Caribbean
3–7	36	9,089	Worldwide
> 7	97	1,609	Africa, Asia, S. America

Source: Derived from UN National Accounts Database.

group confirms the general global pattern. Each of the OECD's 34 member states generates less than 7% of GDP from agriculture with only one exception, Turkey, at 7.5%.

Table 3.1 provides a summary of agriculture and wealth as of 2008.

Wealth and the three industrial sectors

Which of the three industrial sectors relates most strongly to a nation's wealth? Do wealthy countries have more of a focus on manufacturing? Do poor countries have a greater concentration on the services sector? The prevailing view is that manufacturing is the most important sector, and that services indicate low income. But is this view valid? A conflicting impression comes from simply looking at a partial list of services that wealthy countries export: engineering, architecture, medical consulting, finance, business management, air travel, technical training. These are not the low-skilled or low-cost activities that depress a nation's wealth. Quite to the contrary, they show technological refinement, accumulated knowledge, and the buildup of financial capital. Likewise, farming in the wealthier countries has unquestionably shifted from small family operations to powerful large-scale corporations now collectively called agri-business. But it might seem confusing that farming appears to be simultaneously declining and growing. A comparison of each sector to the wealth unravels these contradictory signals.

Agriculture vs. wealth

When viewed as a group, the world's economies show a strong inverse relation of agriculture and GDP per capita (GDPPC). That is, more focus

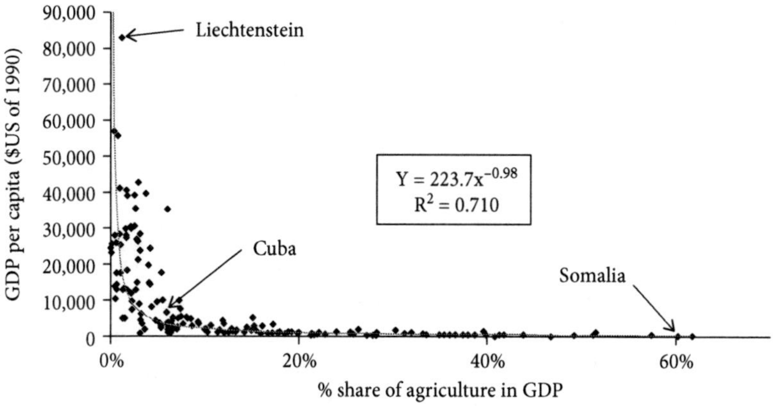

FIGURE 3.3 *Agricultural share vs. GDP per capita: all countries, 2008*
Source: Derived from UN National Accounts Database.

on agriculture is associated with less wealth, and the general pattern stands out very clearly. Figure 3.3 charts each country's agricultural share versus its per capita output in 2008.

Even a quick glance suggests that the group forms a sharply curved, or non-linear pattern. Starting from the lower right hand end of the group, shifting toward the left means shifting from a higher share to a lower share of agriculture. The low level of GDPPC in this area of the chart shows that shifts away from farming are only likely to bring a small change in wealth. So for example, if Somalia were able to make a drastic change such as reducing its 60% agricultural share by half, its national income per capita would rise from $106 to a mere $212 per year.

In the lower left hand corner, the share of agriculture has fallen to less than 10% of the national economy, at which point income starts rising more significantly. Countries such as Cuba have some industrial development including exportable services like tourism. While incomes are still modest, large portions of the population are not mired in deep poverty and starvation. Instead of average incomes in the range of several hundreds of dollars per year, these have risen to a few thousands of dollars.

Finally, the far left hand side of the chart shows a completely different trend. The countries with the lowest levels of agriculture show an extremely sharp increase in wealth. Farming has fallen to less than 3% of these economies, while GDPPC has risen to tens of thousands of dollars

DOI: 10.1057/9781137340313

per year. Liechtenstein is an extreme example but is even surpassed by the tiny Mediterranean state of Monaco, so far off the chart that it has no farming and the world's highest income per person.

Notice that the upper right hand corner of the chart is empty. Absolutely no countries of the world have a combination of high wealth and a high share of agriculture in their economy. This is a key sign that progressing away from the oldest form of industry is deeply associated with rising economic wealth. And it is not merely that high agriculture goes with less wealth. Instead, high agriculture goes with extreme poverty in every case. The trend is confirmed in the lower left hand corner of the chart. Notice again that with under 3% agriculture, there is no extreme poverty as defined by the UN list of Least Developed Countries (see Appendix A for the list).

Focus on the Power law distribution

The data in Figure 3.3 form a hyperbolic or power law distribution. The hyperbolic curve fits the data points well, with a coefficient of determination (R^2) of 0.710. Other curve types can be fit to the data but they have somewhat lower coefficients (e.g. log R^2 = 0.443; exponential R^2 = 0.652) along with the significant disadvantage of deviating from the data at the ends of the range.

This leads to a good theoretical reason to choose a hyperbolic line fit, and that is the asymptotic behavior of the hyperbola at both axes. The scatter plot data in Figure 3.3 indicate that, with the sole exception of Monaco (not shown), the most highly developed economies approach but do not touch the vertical axis. The hyperbolic line fit perfectly captures this feature. At the other end of the scale, many of the least technologically developed economies are the poorest but do not touch the horizontal axis. The transition from low to high income is associated with declining agricultural concentration and increasing marginal returns to technological progress. So, again, the hyperbolic line fit perfectly represents the data trend. Other curves do not support this key theoretical feature and so are not as justifiable as the hyperbola.

Although the power law line fit produces a moderately strong coefficient of determination (R^2 = 0.710), there are two reasons that its equation should not be unhesitatingly used for prediction. First, agriculture is a component of GDP so there will be some element

DOI: 10.1057/9781137340313

of autocorrelation. Second, causation between the two plotted variables is well understood to run in both directions, so either variable could have been arbitrarily chosen as the x-axis, or explanatory variable. In effect, the hyperbola produces the best fit but is nevertheless only likely to be of descriptive value, rather than predictive value. Agricultural share has a clear, moderately strong, and inverse relation with GDP per capita but it should not be used alone in forecasting. The correlation is –0.506.

Manufacturing vs. wealth

Does a high share of manufacturing relate to high wealth? Compared to the clear curve in the agriculture chart above, Figure 3.4 reveals no plain visual pattern from which to form an initial impression. The chart points are widely scattered, or random.[2]

Although there is no discernable relation between manufacturing and national wealth, one interesting pattern stands out in the chart. As was the case with agriculture, none of the wealthy countries has a very high manufacturing share. This shows up as the empty region in the upper right hand corner. The only countries with any semblance of

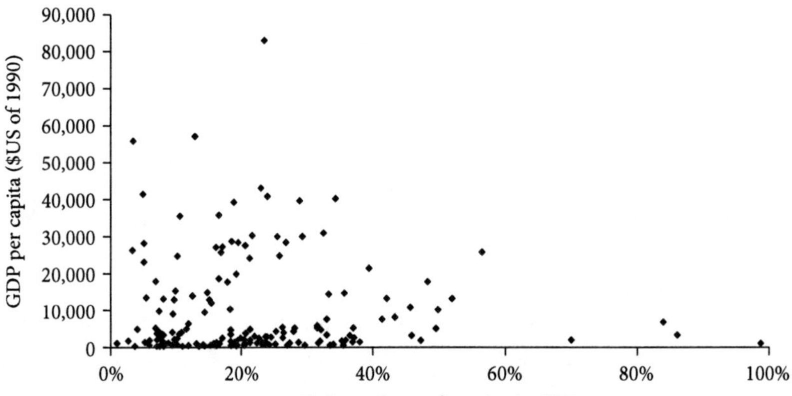

FIGURE 3.4 *Manufacturing share vs. GDP per capita: all countries, 2008*

Source: Derived from UN National Accounts Database.

DOI: 10.1057/9781137340313

high manufacturing and income are small oil-producing states such as Qatar, Brunei, and Kuwait, along with Puerto Rico. These have over 40% in manufacturing and over $10,000 in GDP per capita, but they form a very small minority of states. Most of them also have less factory output than crude oil extraction.

Two further relationships stand out. Countries with low manufacturing shares range from very low to very high income per capita. Countries with low income range from very high to very low manufacturing. So, maintaining a high concentration on manufacturing does not seem to be related to economic wealth.

Current ideas about the relation of manufacturing and wealth seem to be a remnant of the distant past. When Simon Kuznets wrote on this topic in the 1940s, he noticed that the wealthier nations had a strong and rising manufacturing sector. He also concluded that services could not rise to an important degree because they were impossible to trade. After all, how do you export haircuts? At that time, it must have seemed safe to believe that only manufacturing could lead to wealth, and it seems now that the old message had taken firm hold in the minds of policy makers and the public. But today, it has become clear that the fundamental assumption about services was completely wrong; one can indeed export haircuts, or at least architecture, engineering, and financial services. It also seems clear that the old message about industrial structure should be overturned. Maintaining a manufacturing base is still prudent as long as it is technologically advancing, but there should also be a greater recognition of the role of similar advances in the services sector.

Services vs. wealth

The general trend linking the services sector to national wealth is that they rise together. See Figure 3.5. But if the service sector is low skilled, low paid, and of low value, as common wisdom sees it, then why does wealth rise in step with this sector?

More precisely, the chart shows that all of the wealthy economies derive a high share of their GDP from services while an opposing trend also exists simultaneously: many countries with a high concentration in services remain very poor. The trend's direction is clear but it also varies widely, again calling for further scrutiny.

DOI: 10.1057/9781137340313

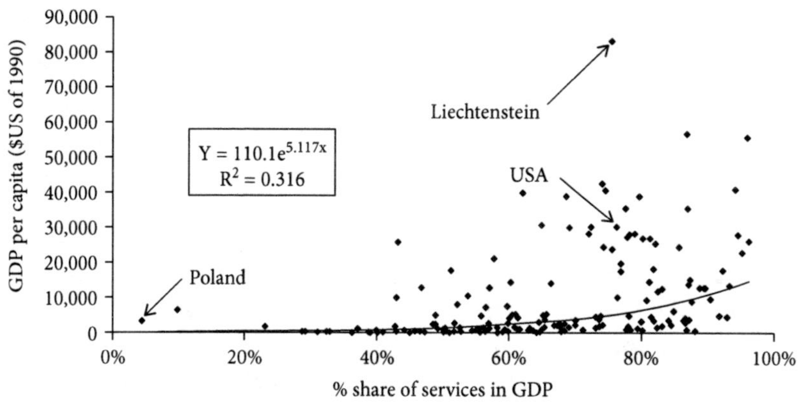

FIGURE 3.5 *Services share vs. GDP per capita: all countries, 2008*
Source: Derived from UN National Accounts Database.

Focus on Best Fit Curve for Figure 3.5

The best fit curve for services data is exponential; R^2 is moderately low at 0.316. The correlation of services and per capita GDP is moderate in strength and positive at 0.427. This indicates that output per person tends to increase as industrial structure shifts toward services, but the correlation is somewhat weaker than with agriculture. The data show a far greater dispersion compared with the agricultural sector above. This is evident as a significantly lower coefficient of determination for services than for agriculture (0.316 vs. 0.710). The conclusion is that the services sector is not a robust indicator of a nation's economic wealth.

The services sector has one unique trait compared with the other two industrial sectors. There are no countries with a very low concentration of services in their economy, as with agriculture at zero to two percent. Even more to the point, no wealthy countries have services below 40% of GDP. This likely indicates that industrial specialization is a precondition of wealth, because services are essentially a form of specialization away from agriculture and manufacturing. But of course, having a high focus on services is not sufficient to assure wealth. This is clear from the fact that at almost any high level of services concentration, income ranges from extreme poverty to wealth. So for example between 60% and 90%

DOI: 10.1057/9781137340313

TABLE 3.2 *Correlation of sector share to wealth*

Agriculture	– 0.506
Manufacturing	+ 0.014
Services	+ 0.427

services, there are very wealthy and very poor countries. The United States and the ultra-wealthy Liechtenstein have the same level of services in their GDP as Senegal, which has 30 times less income.

Extending this comparison can give a clearer understanding of the reasons behind the disparity. While the United States and Senegal have similar shares in the services sector, Senegal's agricultural share remains very high by world standards. Worse yet, its farm productivity remains very low, with 12% of the country's GDP generated by about three quarters of its labor force. In a stark comparison, the United States derives about 2% of its GDP from agriculture but it employs less than 1% of its labor force in this sector. Senegal has not made a substantial shift from farming to manufacturing, so its services sector is not a result of technological advancement, efficiency, and specialization.

To summarize this initial look at the three sectors charted above, only one industrial sector displays a clear pattern of association with a nation's wealth.

Agricultural share has the highest correlation to GDP per capita, but it also has the least dispersed data points in all areas of the chart. In this sense, agriculture clearly represents the economic effect of industrial structure better than the other sectors do. It is not merely a leftover in an age of high technology, or a residual of the manufacturing and services sectors. Indeed, agricultural share is the best prospect to consider as a predictor of national wealth and poverty.

Even as late as the 1950s, there was no correlation between rising wealth and the services sector.[3] Manufacturing was the key to growing prosperity, and it was the rising star of the world's industrialized countries. Of course, the world had yet to see jet aircraft and computers as commonplace. This meant that services were still difficult to trade internationally, and that there was a temporary limit to the growth of this sector.[4] In turn, the lack of trade in services meant that advancements remained local, and several countries lagged in this sector even as their wealth grew from manufacturing. Now that all three sectors are tradable, capabilities have grown more uniform around the world,

DOI: 10.1057/9781137340313

and the resulting correlation has begun to strengthen. That is, a greater share in the services sector indicates a likelihood of greater wealth. As agriculture and manufacturing shares continue declining, it seems that even large countries no longer have a practical limit to their services sector share. This provides further evidence that a new world economy is taking shape.

Notes

1 See UN list of Least Developed Countries at http://www.unohrlls.org/en/ldc /(Accessed February 12, 2013). Also see Appendix A.
2 The correlation is very low at −0.014.
3 S. Kuznets, "Quantitative Aspects of the Economic growth of Nations," *Economic Development and Cultural Change*, Vol. 5, No. 4, July 1957, p. 16; University of Chicago Press.
4 The structure of this argument is from S. Kuznets, op. cit. The limit in the size of the services sector is a result of its non-tradability. If services are largely non-tradable, then they cannot generate export earnings. In order for the services sector to grow to 100% of the economy, food and goods must be imported. But, such imports must be purchased from the proceeds of exports, and if the economy produces only services, then there are no goods to export. So, there is some natural upper limit to the share of services in GDP.

DOI: 10.1057/9781137340313

4
Speeding Up, Slowing Down

Abstract: *Progress is steady and gradual as the agricultural and services sectors switch ranks. Global manufacturing holds its share steady except during the Soviet collapse.*

The rate of change in agriculture correlates to changes in wealth, but predicting future trends based on historic data proves questionable. In a complete reversal from past decades, a falling manufacturing share correlates with rising wealth; the cause is rising productivity. Developing countries can benefit from shifting to manufacturing while advanced countries can benefit by shifting to services. A chart of structural shift can be used as an indicator of political stability. Similarities are revealed between the path of mature industrial economies and today's rising stars. China's current path bears a remarkable resemblance to that of Sweden's of a century ago.

Key words: creative destruction; economic development; economic growth; industrial structure

Atikian, Joe. *Industrial Shift: The Structure of the New World Economy*, New York: Palgrave Macmillan, 2013. DOI: 10.1057/9781137340313.

 DOI: 10.1057/9781137340313

A change of direction

The world's industrial structure has been changing in a familiar pattern for decades. In the 1930s, Austrian economist Joseph Schumpeter famously coined a phrase that describes how industrial economies shift. His term 'creative destruction' captures the very root of what business firms do year after year. They create new products and new ways of doing business that virtually destroy existing firms. Cars replaced horse-drawn buggies just as natural gas replaced household coal and heating oil. CDs replaced cassette tapes just as radios replaced the telegraph. In the 1980s, a British band called the Buggles sang that 'Video killed the radio star', plainly echoing the central concept that new technology brings wealth for some, and ruin for others. These are some of the visible changes, but they also combine to form large-scale industrial shifts. Millions of farm workers have moved to cities and factories, a trend initially lamented as alienating people from the land.

Today's urban dwellers live largely unbothered by never having seen a farm. In recent decades, say since the 1960s, millions of factory workers have left the manufacturing sector and moved to services such as education, health care, engineering, and software development. These service sector workers also seem to go unbothered about never having worked in a factory. But strangely, just as with the erstwhile farmer's lament, the shift away from manufacturing produces a great deal of public angst. Part of the trouble with the supposedly familiar pattern of industrial shift today is that manufacturing has not in fact declined. Although it has moved and shed jobs, manufacturing output continues to grow in almost every region and in the world as a whole.

Economic change always brings conflicting effects for people, countries, and entire regions, as new hardships arise alongside new advantages. The arrival of the automobile decimated the carriage manufacturing industry, with business investment losses and some skilled craft workers rapidly becoming unemployable. The creation of one industry implied the destruction of its predecessor. Benefits shifted from one group to another, in this case due to technological advancement. Upsetting industrial shifts also happen among various geographic locations. In a recent example, the American heavy equipment giant Caterpillar Inc. bought and eventually closed a railway locomotive manufacturing plant in Ontario, Canada. Its operations were moved to the US. A great deal of controversy surrounded the transaction,

DOI: 10.1057/9781137340313

especially as it left several hundred employees without work. The story remained active in the news media for weeks. At about the same time in early 2012, Toyota Motor Corporation announced a planned expansion of a car production plant in virtually the same local area as the shuttered Caterpillar plant. Toyota would add nearly the same number of jobs as was lost in the Caterpillar closing, with two outcomes. First, the total number of local jobs was restored, although there is no doubt that some different workers would get the new jobs. The second outcome, however, was that the news media carried the Toyota story for little more than a weekend. This brief coverage helps to cement the popular and wrong impression that unemployment usually goes in only one direction: worse.

Can the overall effect of changes in technology and changes in location somehow be correctly gauged? Consider two simple questions. If farms have been in decline since 1900, why has overall employment not collapsed? If factories have been in decline since the 1960s, why has unemployment held relatively steady? These two declining trends have been in action for so long in the dominant Western economies that if they were real, no jobs would remain today. Instead, as the first two industrial sectors continue expanding their output, the services sector grows even faster. This makes it seem that agriculture and manufacturing are declining even though they are not. Relative size is indeed changing, but this is much harder for a casual observer to perceive correctly.

Another aspect of the shifting industrial pattern shows up in country comparisons. Agriculture can rise or fall; wealth can rise or fall. Figure 4.1 shows each country as a point representing its agriculture and wealth. It may seem murky at first but the pattern is simple: the cluster of points is centered along the diagonal line. One combination dominates. From the center of the chart, follow the diagonal line upward and to the left: agriculture falls while wealth rises. This kind of shift happens in about three quarters of all countries, and is usually accompanied by political stability and improvements in the quality of life. Think of it as the general direction of industrial shift.

The most troubled countries usually shift in the opposite direction along the diagonal line: agriculture raises its share in the economy while income per person declines. Such cases make up fewer than 10% of the world's countries, where rising agriculture can be a sign of civil war that destroys industries while leaving far flung farms somewhat less disrupted.

DOI: 10.1057/9781137340313

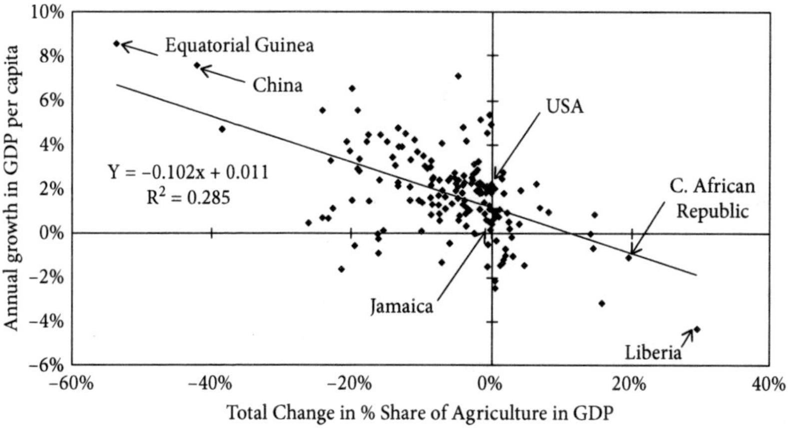

FIGURE 4.1 *Velocity of agricultural share vs. GDP per capita, all countries*
Source: Derived from UN National Accounts Database.

The worst off instances are Liberia and Central African Republic where annual income per person has fallen over nearly forty years. Meanwhile, the share of agriculture in their economies has risen by more than 20% over the same period, a hugely regressive outcome. In these rare and desperate instances, farm output falls but the other sectors fall even further. This is de-industrialization through war and societal collapse.

Very few countries, less than 10% of the world, experience rising agriculture combined with rising wealth. About half of these are small developing countries. Chile is a larger example that has growing farm exports and steadily rising wealth, along with a vastly improved political climate over the past four decades. Other larger members of this group are Iran, Netherlands, Nigeria, Saudi Arabia, and UAE, all of which are significant oil producers. This shows that they have benefited from the large increase in crude oil prices since 1970, and have been able to increase agricultural investment to their advantage.

In the fourth and final direction of change, agriculture and income fall together over the course of decades. Only ten countries lie in this unfortunate state: Burundi, Côte d'Ivoire, Gambia, Guinea-Bissau, Haiti, New Caledonia, Palau, Qatar, Somalia, Zimbabwe. Most suffer industrial destruction from political–military conflict, or natural disasters such as flood, drought, hurricane, or earthquake. Six of these countries are in a single continent, Africa.

DOI: 10.1057/9781137340313

The pace of change

While the shift from farm to factory to office is widely recognized, the pace of that shift and its effects remain somewhat mysterious. So which countries are shifting and by how much? How does growth relate to changes in structure? Do the rapidly changing structures generate more wealth than slowly changing ones?

It might seem reasonable to think that slow and steady change would be better, or easier to deal with, than rapid change. That is not always true. Around the world, the pace of change within countries and regions varies so much that it might appear chaotic. There is however a clear pattern tying industrial structure and economic growth closely together. Part of that pattern can be easily seen at the global level. The total of all the world's economies is shifting gradually away from agriculture. Agricultural output continues growing, but its share in world GDP is shrinking.

The pace of industrial shift has a peculiar relationship with economic growth. In most cases, the faster a country's agricultural share declines, the faster its wealth grows. Unfortunately, the opposite is also true. Faster *increases* in the agricultural share come with faster declines in wealth. This can be seen at the far right end of the chart in Figure 4.1. Of the world's ten fastest shrinking economies (Liberia, DRC, Djibouti, Kuwait, Palau, Zimbabwe, Niger, Nauru, Somalia, Togo), six are small African nations. Of the world's ten fastest growing farm sectors (Liberia, CAR, DRC, Guyana, Sierra Leone, Comoros, Benin, Iran, Tuvalu, Niger), again, six are small African nations. The conclusion must be that a growing farm sector is generally not a good economic sign.

The world's fastest growing economies are China and Equatorial Guinea on Africa's west coast. They are also undergoing the world's fastest shift away from agriculture, although they remain two vastly different cases. While China's industry has shifted from farming to manufacturing, the miniscule Eq. Guinea has shifted into oil extraction, recently rising to the third-largest producer in its region. Crude oil remains one of the most disruptive commodities for small countries because export earnings can easily be concentrated in a very small number of hands. In Eq. Guinea's case, such a result has led to far less distributed benefits from industrialization than in China's case. Nevertheless, these two countries confirm the widely known fact that when lower income countries grow quickly, they grow more quickly than any wealthy country can.

DOI: 10.1057/9781137340313

The general connection made clear here is between the speed of change and the speed of growth. This rapid change can also bring worries of mass displacement, difficult adjustment, or even civil unrest, but this does not always happen. South Korea, for example, has made exactly this kind of change with no more hardship than many slow growing countries. It has seen labor disruptions and protests, but these are not uncommon in most countries regardless of changes in their industrial structure. The problems of very slow growth, namely stagnation, could be even more difficult than dealing with rapid change. In a stark contrast with several of their regional neighbors, Jamaica and Bolivia have seen virtually no industrial shift and no economic growth over the past forty years. Even without widespread civil strife, such a long stagnant period cannot have produced anything but dwindling prospects.

While low income economies are always among the quickest growing, highly advanced and wealthy economies always grow slowly. This kind of slow growth is not stagnation, but rather a steady, almost unbroken pace lasting decades and even centuries. The effect of slow steady growth eventually appears as tremendous wealth, several times greater than economies with erratic growth. In every such case, agricultural share has declined all along the way, quickly at first, but gradually slowing its pace. Where low income China saw an increasingly rapid pace of economic growth in recent decades, the US saw virtually no change in its growth rate. And where China's agricultural share has rapidly shrunk from about 50% to 10%, the US share has stopped shrinking at all, stabilizing at 2% of GDP. Wealth and agriculture continue speeding up and slowing down, but the pace differs depending on the state of the nation's economy. The general pattern remains simple: all of the wealthiest nations change gradually, while the poorer nations range from rapid self destruction to stagnation to the most rapid advance.

A more detailed view of a few individual countries shows how much variance there can be in structural shifts. Figure 4.2 shows Uganda's agricultural concentration having fallen by one half in just 38 years, while in contrast, the United States has fully stabilized. Of course the US has already undergone the same shift as Uganda at a much earlier time. This is the predominant pattern all over the world, with three quarters of all countries exhibiting a decline in agricultural concentration.

Some countries counter the global trend, ending the period with a higher agricultural share than at the start. Figure 4.3 charts the trend in industrial structure for all major countries with this trend.

DOI: 10.1057/9781137340313

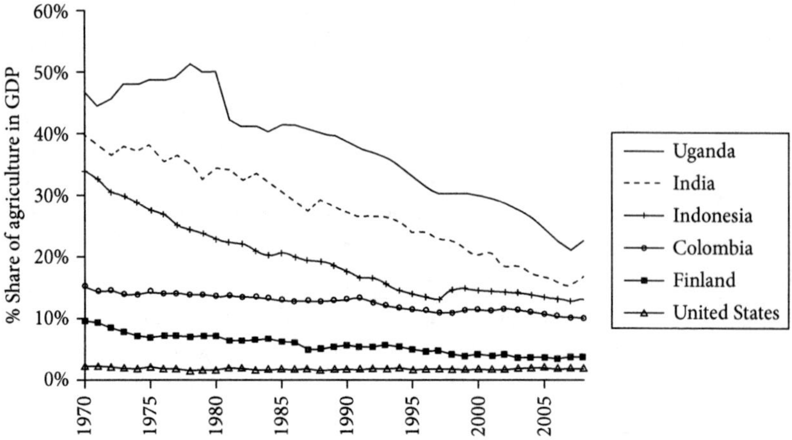

FIGURE 4.2 *Selected countries with a falling agricultural share, 1970–2008*
Source: Derived from UN National Accounts Database.

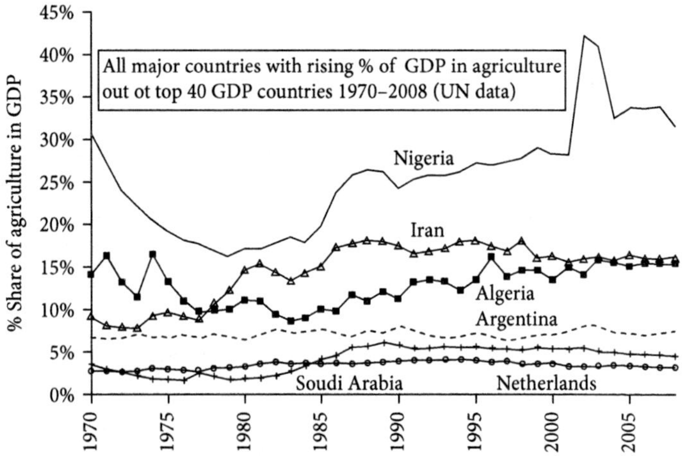

FIGURE 4.3 *Selected countries with a rising agricultural share, 1970–2008*
Source: Derived from UN National Accounts Database.

Even within this top-40 group of large economies, greater farming concentration is consistently associated with poorer countries. For example, Nigeria ends 2008 with the highest share of GDP in agriculture and the lowest per capita GDP ($737). The Netherlands has the highest output in the group at $28,415 and it has the lowest agricultural concentration.

Over the 38 years since 1970, countries with a faster decline in the share of agriculture tend to have faster growth in output per capita. A rising share

DOI: 10.1057/9781137340313

of agriculture is seen in the minority of countries, mainly comprising the poorest countries and several oil-producing states. Of course a faster rise in agricultural share also correlates with a worsening output per capita. But it should also be made clear that some of the worst off economies are not benignly shifting or reverting to agriculture. Rather it is often a severe military conflict such as protracted civil war that destroys a significant amount of industrial capacity, thus leaving agriculture at a higher proportion of GDP. An additional factor that distorts the picture is that GDP appears to decline only modestly. In fact, foreign aid and family remittances partly compensate the large losses caused by physical destruction.

National stability and structural shift

Changes in industrial structure need not carry on smoothly year after year. Argentina shows just how unstable an advanced industrial economy can be. Over a span of about five years, its agricultural share can jag between 6% and 8%, and back again. While this may seem like an insignificant amount, it lies on the threshold between the world's poor countries and those in transition to the high income ranks. It also suggests that the country has made little if any technological progress. Furthermore, Argentina's ruling political parties change hands year to year. This means that import–export regulations change with each administration, as do the basic approach to labor, wealth distribution, monetary policy, and foreign debt. So the country's wealth also lurches back-and-forth in synchrony with its fluctuating industrial structure.

Argentina carries a kind of mystique for economists, always poised to break through from the middle ranks to truly wealthy status, and yet always delivering disappointment. Argentina's jagged pattern of agricultural shift mirrors its great potential and middling results.

In stark contrast, South Korea's agricultural share carried on declining smoothly for decades. Comparing the total change in wealth in the two countries is nearly shocking. Over the 38-year period, Argentina's GDP per capita rose from about $5,000 to $7,500. Meanwhile, South Korea's rose from about $2,000 to over $14,000. Stability seems to count for much more than just political tranquility.

The chart of each country's structural shift reflects the degree of continuity or smoothness in its economic development. Argentina's path in Figure 4.4 displays a particularly unstable economy. Its GDP composition

DOI: 10.1057/9781137340313

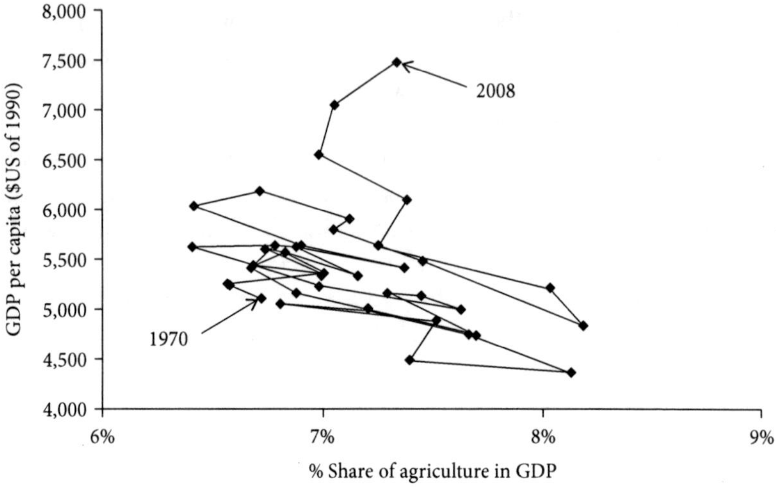

FIGURE 4.4 *Argentina agricultural share vs. GDP per capita, 1970–2008*
Source: Derived from UN National Accounts Database.

repeatedly swings between higher and lower agricultural concentration while its output level rises and falls. In some years, agriculture and output rise together, while in other years, they veer in opposite directions. There is no clear trend in the economy's direction, and its total growth over nearly four decades is modest. These could be signals of national scale ailments such as vacillating economic policy or social unrest. Erratic curves like Argentina's can give contradictory signals depending on the time segment being observed. For example, the full 38-year plot indicates rising output, while a shorter time span like the 1980s shows declining output.

South Korea's chart bears no resemblance to Argentina's. Nor does its economic growth record. Even a quick look at South Korea's path in Figure 4.5 suggests an economy developing in a clearly defined manner.

The direction and spacing of the points show little yearly change throughout the period. GDP per capita increases gradually in most years, as agriculture's share declines in fractions of a percentage point. As its farm sector declined in importance from 25% to 5%, South Korea's manufacturing has leaped to the foreground with advances in heavy machinery, semiconductors, automobiles, and consumer electronics. With Argentina's essentially stagnant industrial structure, one would be hard pressed to list any such well-known advancements.

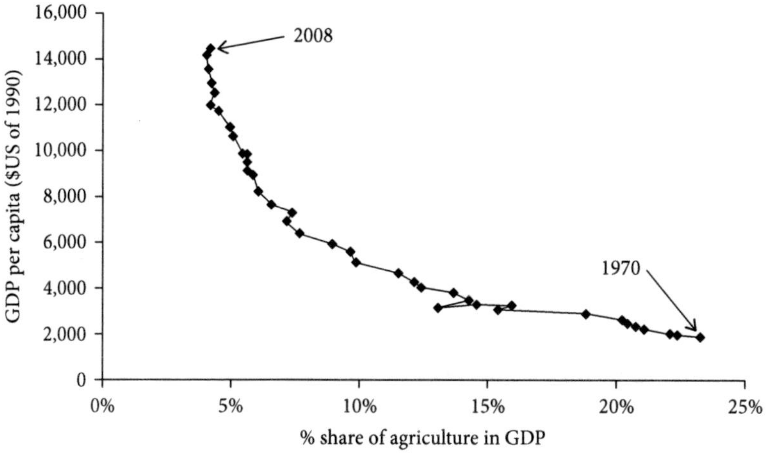

FIGURE 4.5 *South Korea agricultural share vs. GDP per capita, 1970–2008*
Source: Derived from UN National Accounts Database.

Argentina and South Korea exhibit two basic types of industrial transition that may simply be called rough or smooth. Likewise, their societies may be thought of as being chaotic or orderly to varying degrees. Another chaotic example is war torn Sudan, while some orderly charts come from China, America, and Sweden. France and Laos lie somewhere in between these extremes, showing a path that cuts back-and-forth, while also trending in a clear direction. The chart for any country provides a lot of detailed information, but it can also be an easy way to assess a country's economy over time, as it provides a sense of progress, growth, and economic stability at a single glance.

A world of change

A quick glance at Figure 4.6 gives the true impression of a gradual shift in each main industrial sector. The global share of agriculture slowly falls while manufacturing holds nearly steady, and services rise. Underlying this trend, agricultural productivity steadily rises. Progress and gradualism characterize these shifts in the world's economic structure.

Less well known is that real output in manufacturing and agriculture has only ever declined for the five years following the convulsive end of the Soviet era. Barring this catastrophe in which the communist sphere

DOI: 10.1057/9781137340313

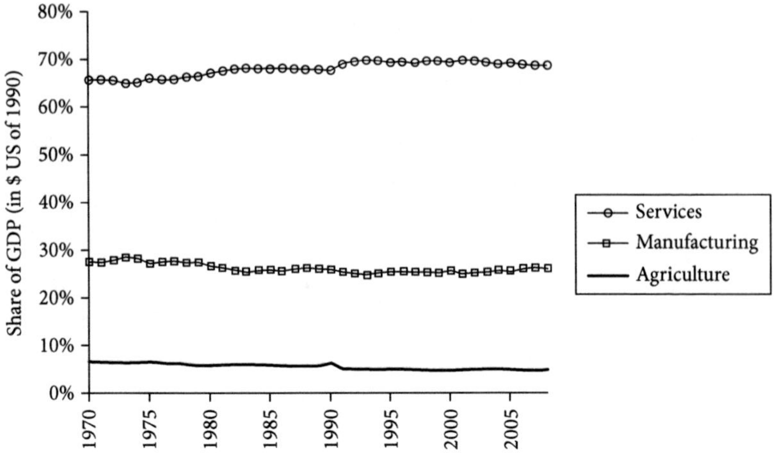

FIGURE 4.6 *Three sector trend in global industrial structure, 1970–2008*
Source: Derived from UN National Accounts Database.

lost three quarters of its factory output, world manufacturing would have maintained a stable share instead of a slight decline. Contrary to the worry about the death of manufacturing, global factory output grew by 200% from 1970 to 2008. Better yet, the global growth rate of farm output consistently surpassed population growth.[1]

Wealth rises as the share of agriculture falls. Figure 4.7 traces the annual shift of industrial structure relative to GDP per capita at the broadest level. Each point on the graph is the annual average of all countries.

The global pace amounts to an average gain of $1,400 per capita for every 1% drop in the share of agriculture over the period. Overall, a shift of 1.7% in agriculture over the course of four decades may seem to be insignificant, but it is associated with enough technological progress to raise global GDP per capita by nearly 80%.

The world's agricultural output did not drop from 1970 to 2008. In fact, it rose by 140% to nearly $2 trillion per year. During the same period though, world population increased by only 83%, meaning that farm output far outpaced population growth. Figure 4.8 charts the world's farm output through the period. It shows that total output grew continuously through the period, except immediately following the 1989 collapse of the former Soviet Union. Even with this important regional decline, annual output recovered and ended the period at more than double its initial level (see the left hand scale of the chart).

DOI: 10.1057/9781137340313

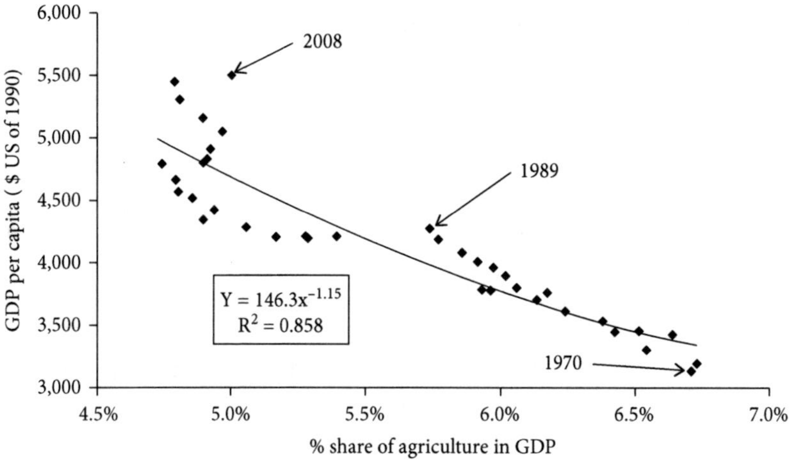

FIGURE 4.7 *Trend in global agricultural share vs. GDP, 1970–2008*

Source: Derived from UN National Accounts Database.

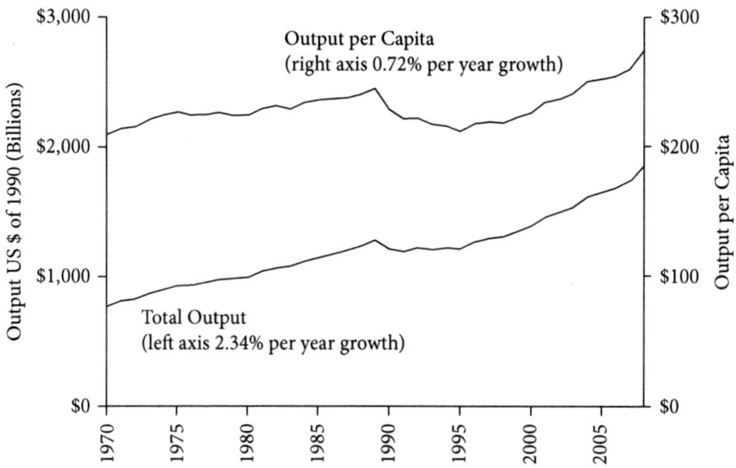

FIGURE 4.8 *Trend in world agricultural output and output per capita, 1970–2008*

Source: Derived from UN National Accounts Database.

The chart also shows agricultural output per capita, which is crucial because population growth can eradicate the gains made by farmers. The per capita trend follows the total output trend, which suggests that more food was available per person through the period. Of course these global figures are only averages, so there is always the possibility that some regions or countries would not benefit from the rise in farm production.

In light of ongoing concerns about looming worldwide food shortages, it may seem controversial to report that agriculture is growing at a higher rate than the human population. A second credible data source confirms the result. The UN-FAO Summary of Food and Agricultural Statistics[2] reports sectoral growth rates over the 1970s, 1980s, and 1990s. Growth in production held steady over these three decades at 2.3% per year. The corresponding population growth rate remained lower at 1.5% in the 1980s and 1990s (no data were reported for the 1970s). The same pattern held true for Africa and Asia, albeit by a smaller margin. By this measure, world agricultural output grew about 50% faster than the population for every region worldwide. The composition of this output has slightly changed as the developing countries began shifting their consumption from grains to meats, but food availability in calories per capita has grown in every continent over the past four decades.[3]

Can the wealthy nations feed themselves without focusing on agriculture, and if not, do they contribute to the impoverishment of other nations by importing their farm products? The answer is that the wealthy countries do feed themselves. They still have a very robust agricultural sector, and even export large quantities of food. For example, the US exported nearly $140 billion in agricultural products in 2011, while importing only $107 billion.[4]

US agriculture comprises just 2% of its GDP while India's is 17%. Nevertheless, India, with over quadruple the US population, has 18% less real agricultural output. In 2008 India produced $167 billion next to America's $197 billion in this sector. In general, wealthier countries produce more agricultural output than poorer countries of the same size. Some confusion may stem from the less visible fact that wealthier countries simply have a far greater output in services, which reduces their proportional share of agricultural output. Again, a low share of agriculture in the GDP of advanced economies is almost always the result of rapid growth in manufacturing and services, and not a real decline in the amount of farming activity.

Focus on agricultural output

Of the more than 200 countries listed by the UN, only two major economies, the former USSR and Japan, have had a real decline

DOI: 10.1057/9781137340313

in agricultural output since 1970. The Soviet Union's yearly farm output experienced a disastrous plunge of over 50% during the 1990 collapse, from which its successor states have not yet recovered. This event was the sole shock to the global system since 1970, erasing over 20% of accumulated global agricultural growth. Meanwhile Japan saw a very gradual but continuous decline of 0.3% per year over the past 40 years, erasing a mere 0.8% of world growth in the sector. A handful of minor economies, mostly small tropical islands, have also seen a decline in real output but combined they account for a negligible 0.7% of global agricultural output. (These are Switzerland, Greece, Singapore, Luxembourg, Trinidad, Cyprus, New Caledonia, Haiti, Barbados, San Marino, Liberia, Palau, Kiribati, Montserrat, and Nauru.)

Contrary to the idea of agriculture declining due to a shifting industrial structure, annual real agricultural output per capita in the US has increased from $341 to $649 since 1970, an average compounded growth of 2.76% per year. The sector is growing faster than the 1% population growth, at nearly same 3% rate as the national economy, and so is clearly not in decline. Sweden presents a confirming example with its sector output expanding from $826 to $1015 per capita, or 0.54% per year.

In aggregate, the global agricultural output trend per capita remains consistently positive with an average annual increase of 0.72% since 1970, equivalent to 31% real total growth. Table 4.1 shows the decomposition of per capita values into sector growth and population growth. Not only has the world's agricultural output significantly advanced, but its growth was greater than population growth and was evident in the vast majority of countries.

TABLE 4.1 *Global agricultural output growth, 1970–2008*

	Total growth (%)	Annual average growth (%)
Farm output	141	2.34
Population	83	1.61
Farm output per capita	31	0.72

Source: Derived from UN National Accounts Database.

~ ~ ~ ~

DOI: 10.1057/9781137340313

At the dawn of the First World War, the manufacturing nations stood as the world's powers. These were Western Europe, UK, and Scandinavia, with over 30% of their GDP in this sector. America had just become the world's largest economy in 1906 and would soon join the group of major manufacturers. Meanwhile Japan, China, and Korea remained in the intensively agrarian background. Whoever shifted from farms to factories most quickly saw their income and power rise, a trend that remained true only until the 1960s but remains wrongly stuck in the common wisdom until today. Instead, the real economic powers since 1970 were those that maintained farm and factory output while rapidly growing their services sector. Whereas nearly all countries had a similar level of services in the early 20th century,[5] today's wealthiest countries are distinguished by having the heaviest concentration in services.

Until the 1960s, it must have seemed reasonable to assume that the large industrial economies had an upper limit on the size of their service sector. It likely seemed just as reasonable to assume that manufacturing would forever be the crucial activity for generating wealth. After all, something exportable was needed to pay for imported products, and exporting services in those days was prohibitively expensive because people would need to commute overseas in order to deliver those services. Electronic communication and jet travel have either eliminated commuting, or slashed its cost. Any upper limit on the services sector has faded away, and the modern anxiety about manufacturing should now fade away too.

Future shift—an estimate

Can industrial structure be used to predict economic growth? Can today's data be used to predict how long it might take for the world economy to reach some specific level of income? The chart in Figure 4.7 suggests a preliminary answer. The trend is that wealth rises as the agricultural share continues falling. Using the equation from the chart, one estimate is that it could take 44 years for the agricultural share to fall another 3%, at which time the world's GDP per capita would have doubled.

Focus on prediction: Figure 4.7

A linear equation can be fitted to the data as $y = -85{,}543x + 8{,}980$. The line fit is good with a R^2 value of 0.830, very nearly as good

DOI: 10.1057/9781137340313

as the hyperbolic curve shown in the figure. Projecting to a world agricultural share of 2% gives a GDPPC of $7,269. The difference in the GDPPC estimates between the two curve fits is very substantial though. This shows the importance of choosing the curve type not just by mathematical fit, but also by the concept behind it. The linear curve fit is clearly the wrong choice because it generates an absurd prediction: extending the line toward the right hand side of the chart results in zero income when the agricultural share reaches about 7%.

The trend line equation from Figure 4.7 is $y = 146.3x^{-1.15}$. This expresses world aggregate GDP per capita growth as a power function of agricultural share. Projecting with a world agriculture concentration of 2%, the corresponding per capita GDP is $13,154 or more than double the actual 2008 output level. The compound average rate of change per year has the same form as the power function in Figure 4.7, and can also be used to project the time for agriculture's share in GDP to further decline from 4.8% to 2%. This is calculated to take 44 years, although the uncertainty inherent in using the power function for prediction is typically extremely high.

Figure 4.9 shows the change in world GDP per capita as industrial structure shifts. The concept is straightforward.

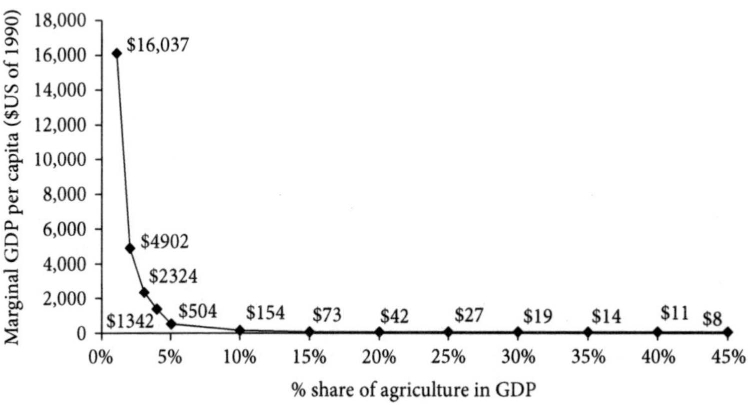

FIGURE 4.9 *Rise in global GDP per capita as agricultural share falls*

Source: Derived from UN National Accounts Database.

DOI: 10.1057/9781137340313

In the average poor country with a high concentration on agriculture, income rises only slowly. This shows up as the shallow sloped line starting in the lower right hand corner. Following the line toward the left means that agriculture is declining in importance. Eventually income grows very rapidly as the line turns upward. The dollar values shown along the line show how much the income per person rises for each percent that agriculture falls. As agriculture falls below 5% of GDP, each percent drop comes with thousands more in income per person. Perhaps the clearest example currently underway involves China's breakneck shift from farming to manufacturing since regaining Most Favored Nation status by the US in 1980, and acceptance to the World Trade Organization in 2001.

Focus on marginal GDP per capita

Read the graph in Figure 4.9 in the following way:

For any point (x, y) on the curve, a 1% decline in the global share of agriculture to x% is associated with a rise of $y in global GDP per capita.

For example, see point (x, y) = (10%, $154). A 1% decline in agriculture's share in GDP is associated with a rise of $154 in global GDP per capita.

These values of marginal GDP per capita were calculated from the equation in Figure 4.7, the global aggregate, in which the sequence of annual data points may give the impression of gradual change. But, while gradualism is true of the structural shift, marginal values help to underscore the varying intensity of the relationship between structure and output. With a high concentration of agriculture, as many poor countries have, a small shift in structure is associated with a small increment in output. But as economies develop, the impact on output becomes explosive.

This structural shift can be expressed as a difference, or as a proportion in the following example. Either method gives the same kind of conclusion. From 1970 to 2008 China's agriculture dropped from 53% of GDP to 10% of GDP. This –43% difference came with a per capita output gain of $1,400. Meanwhile South Korea's agriculture fell from 22% to 4%, a difference of –18%, with an output gain of $11,600. South Korea's smaller drop in agricultural share saw a much greater rise in per capita output than China's. This increasingly steep rise in per capita output is typical of most economies

DOI: 10.1057/9781137340313

as they continue the transition away from agriculture. That is, marginal per capita output rises as agricultural concentration falls. Again, Figure 4.9 shows this relation in graphical form.

Expressed as a proportion, China's agricultural concentration fell by 81.1%. In the same period, South Korea's fell by a nearly identical 81.8%. Expressed in this alternative way, the same degree of change in industrial structure, 81%, came with a far greater output gain for South Korea, in fact 8.3 times greater than China's output gain. Less agricultural concentration is not only associated with a higher level of total output per capita, but also with a higher marginal output.

Early writers in this field commonly used correlations between the main industrial sectors and GDP as a tool to analyze growth. In his 1957 paper, Simon Kuznets found that a rising share in manufacturing was correlated with rising economic wealth. In other words, Simon said that wealth goes as manufacturing goes.[6] By doing so, Kuznets may have laid the foundation for today's lament about the supposed decline of manufacturing. People still worry about the declining share in manufacturing, unaware that its real output is still rising, and that technology has dispensed with the 1950s era of non-tradable services. More striking still is that the correlation between manufacturing and income has been eradicated. This shows that national economic development has fundamentally changed in the past few decades. Development was once associated with a rising share of manufacturing. Now it is associated with a declining share of manufacturing in the most advanced economies.

Predictable uncertainty

As the agricultural share falls, wealth suddenly changes from rising gradually to rising rapidly. This sudden change in speed makes mathematical predictions erratic and unreliable. Is there another way? Perhaps estimates can be made using longer term industrial trends than those in the UN database that start in 1970. Similar trends can be compared from the 19th century, when today's advanced economies were still developing their farming and manufacturing sectors. That could boost confidence in using today's data to make predictions about future shifts.

The US Bureau of Economic Analysis and the Census Bureau have some relevant data reaching back to 1947. Simon Kuznets provides two further

DOI: 10.1057/9781137340313

data points that extend this series back to 1869. Combining these with current UN data creates a 140-year US history. Sweden already has a similar history available starting in 1870,[7] which may confirm the US case.

First, Sweden's historical data are used to gauge the accuracy of prediction in general. Then the long course of development in Sweden and the US shed light on China's recent history.

See Figures 4.10 and 4.11.

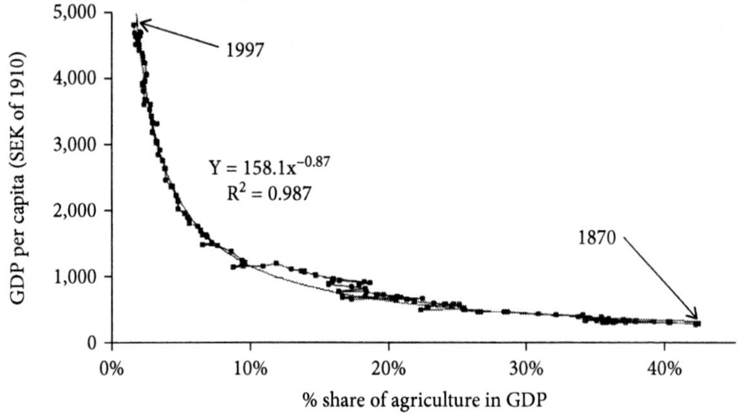

FIGURE 4.10 *Sweden* historic *agricultural share vs. GDP per capita, 1870–1997*
Source: Derived from GGDC database.

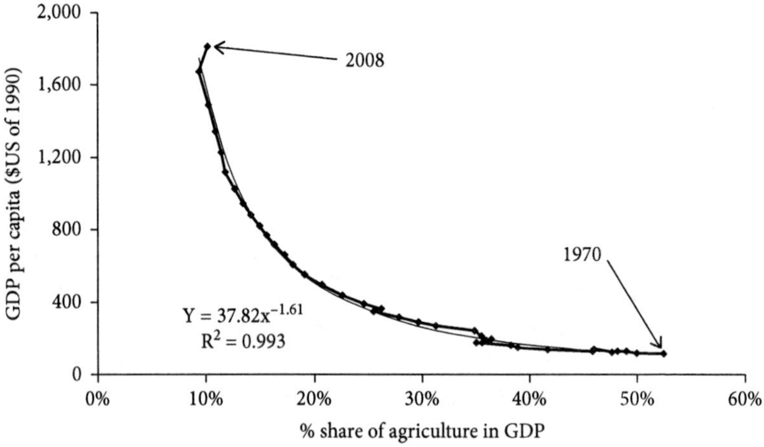

FIGURE 4.11 *China* recent *agricultural share vs. GDP per capita, 1970–2008*
Source: Derived from UN National Accounts Database.

DOI: 10.1057/9781137340313

Sweden's development history from 1870 is similar to China's starting from a century later, 1970. They started at those times with similar industrial structure, about 40% and 50% in agriculture, respectively. These values then declined in the familiar pattern as income increased in both countries. Could Sweden's longer history be used to predict China's coming development? This could be done by looking at the first half of Sweden's data, and using it to predict the second half of Sweden's data. If there is a close match in Sweden's case, then China's recent past may be used to forecast its near term future.

This method was used with Figure 4.10. A new equation[8] was generated for Sweden from 1870 to 1908, the same length of time for which data on China is available. Using this equation, Sweden's GDP per capita was predicted to be 8,133 when its agricultural share reached 2%. The actual historical outcome at the 2% level was 4,425 or nearly half the predicted value. The margin of error is nearly double, which is by far not accurate enough a result to recommend this method for prediction.

This example shows the risk of making predictions while ignoring known limitations, in this case, the shape of the curve on the chart. Each country's chart shows a horizontal and vertical segment in its curve, and the equation treats these as symmetrical. The country's industrial development over time need not be symmetrical though, so the prediction will more than likely be inaccurate. Nevertheless, the general pattern holds true in the majority of cases; it still seems reasonable to expect industrial economies to grow as their agricultural share falls.

Over the course of about 140 years, America's industrial shift has been very slow compared to China's recent decades. See Figure 4.12.

The US farm share of the economy went from 20% to 2% in that 140-year period, whereas China went from 50% to 10% in only 35 years. China had made a much larger change in a much shorter time span. What about the time it took to progress through similar agricultural stages? The US decline from 20% to 10% agriculture took about 60 years, whereas it took China a mere 15 years. It seems clear that from either perspective, China has developed much more quickly than America, although this might be illusory.

These two countries obviously present very different cases. The level of technology available at the time of the transition was a great influence. When the US made its shift from 20%, it already had most of the technology available anywhere on earth such as railways, telegraph, steam engines, and so on. This means that the US could not accelerate its

DOI: 10.1057/9781137340313

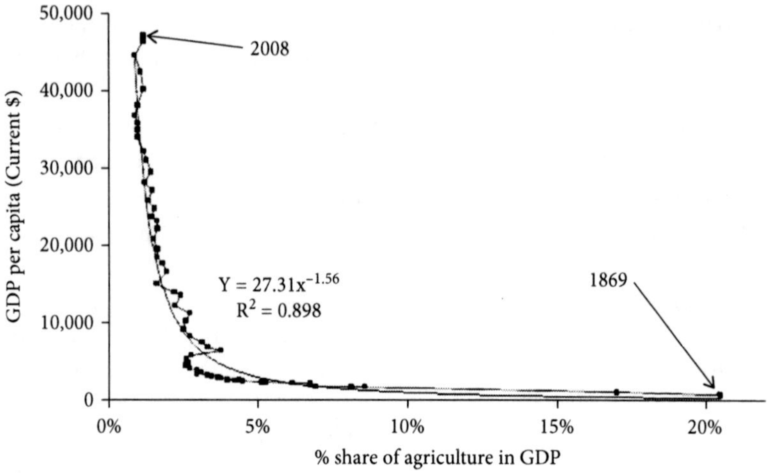

FIGURE 4.12 *US* historic *agricultural share vs. GDP per capita, 1869–2008*
Source: Derived from S. Kuznets, UN National Accounts Database, US Bureau of Economic Analysis, and US Census.

shift by importing a vast array of new technology such as transportation equipment or machinery. It had to become a business innovator, a far slower process. In contrast, China in the 1970s was far behind the rest of the advanced industrial countries, so it had the opportunity to import technologies that would enable rapid change. Perhaps the other main difference stems from the speed of political decisions. While policy decisions that affect industry tend to be incremental in Western democracies, modern China can enact such decisions almost instantaneously by comparison. Once a decision is taken, the resources of the entire nation can be directed at rapidly building major new cities and an industrial infrastructure that draw down the farming sector.

Comparing the three country graphs above plainly shows that the patterns were the same in each case. Agricultural share fell as income rose, slowly at first and then increasingly more rapidly. The fact that each country had a vastly different makeup suggests that the industrial development path does not closely depend on these factors. In history, America was born of a colonial rebellion while ancient China grew by assimilating its neighbors, and Sweden built itself up by sea trade and conquest. The three countries' locations around the globe could hardly be more diverse, nor their languages more unrecognizable to each other.

DOI: 10.1057/9781137340313

Perhaps most striking, though, is that their founding political systems cover the entire range from monarchy to democratic republic to single party dictatorship. The only similarity is that none of these countries is tropical; instead each is situated in a temperate seasonal climate zone.

Figures 4.10, 4.11, and 4.12 show a remarkable similarity of form between US, Swedish, and Chinese structural shifts, even at greatly varying timescales. All of these economic parallels across the world suggest that when poorer nations embark on an industrial path to development today, they face a strong likelihood of attaining similar improving results.

Notes

1 http://data.worldbank.org/indicator/SP.POP.GROW/countries/1W?display=graph (Accessed January 11, 2013).

2 Food and Agricultural Organization of the United Nations, *Summary of Food and Agricultural Statistics 2003*, Rome, 2003; p. 69.

3 Ibid., p. 72.

4 U.S. Department of Commerce, Bureau of Economic Analysis, International Accounts Products for Detailed Goods Trade Data. *U.S. Trade in Goods (IDS-0008)* http://www.bea.gov/international/detailed_trade_data.htm (Accessed January 6, 2013).

5 Ibid., pp. 10, 18, 19.

6 S. Kuznets, "Quantitative Aspects of the Economic Growth of Nations: II. Industrial Distribution of National Product and Labor Force," *Economic Development and Cultural Change*, Vol. 5, No. 4, Supplement (July 1957), p. 9.

7 J. P. Smits, P. J. Woltjer and D. Ma (2009), *A Dataset on Comparative Historical National Accounts, ca. 1870–1950: A Time-Series Perspective*, Groningen Growth and Development Centre Research Memorandum GD-107, Groningen: University of Groningen. http://www.rug.nl/research/ggdc/data/historical-national-accounts (Accessed 12 February 2013).

8 The equation is $y = 114.4x^{\wedge} - 1.09$.

DOI: 10.1057/9781137340313

5
Power Shift: East and West

Abstract: *Most countries including China are losing agricultural share in GDP, even as food production grows faster than population. The global manufacturing share has held steady over recent decades, but it has declined in Western countries and risen in Eastern countries. China is, of course, a large part of the story, but so too is Japan's stagnation, Korea's rise, and Europe's drop in the ranks. The global pie has grown and so has the Eastern share, while the decline of Western manufacturing is greatly exaggerated.*

Tradable services constitute the current revolution in the global industrial shift.

Key words: China pivot; East–West; new world economy; structural shift; trade sector

Atikian, Joe. *Industrial Shift: The Structure of the New World Economy*, New York: Palgrave Macmillan, 2013.
DOI: 10.1057/9781137340313.

 DOI: 10.1057/9781137340313

Industry, wealth, and power

Economic wealth brings freedom, choice, power, and prestige to the individuals that hold it. For nations, wealth brings all of this and security, or the ability to defend, repel, and overtake. The key distinction among nations, though, lies in the combination of technology and size.

National power encompasses far more than military might. As 1939 approached, Germany was feared not only for the size of its army but also for its productive capacity that turned science into mass technology. Whereas the Napoleon of the previous century believed that an army fights on its stomach, the new force fights on its machines. Farming to feed soldiers had been overtaken by manufacturing to mechanize battles.

The same shift happened in the daily life of nations. Power is not often asserted in massive battles, but in an ongoing stream of economic decisions. Powerful nations influence global trade patterns. They decide the legal framework and the dispute mechanisms. They organize the world's financial structure, and sway the environmental agenda. And the powerful nations are those with wealth that flows primarily from industrial progress. Large, powerful agrarian nations no longer exist. Instead as the world's industrial shift continues, more large and powerful manufacturing nations are emerging, and many of the leaders are already focusing increasingly on services.

Industrial shifts are always sectoral but they can also be geographic, with economic power shifting from one country or region to another. The broadest geographic pattern today shows a relative decline in agriculture all over the world simultaneous with the flow of industrial power from West to East based on growth in manufacturing. One of the most persistent and important global stories in recent years is the resurgence of China and the East in general as an industrial powerhouse.

Mapping the agricultural shift

Figure 5.1 shows a ranking of countries according to the amount of change in their industrial structure.[1]

The striking feature is that most countries are changing, and doing so by a significant amount. About three quarters of the countries have a falling share of agriculture in GDP, confirming that this is a global shift

DOI: 10.1057/9781137340313

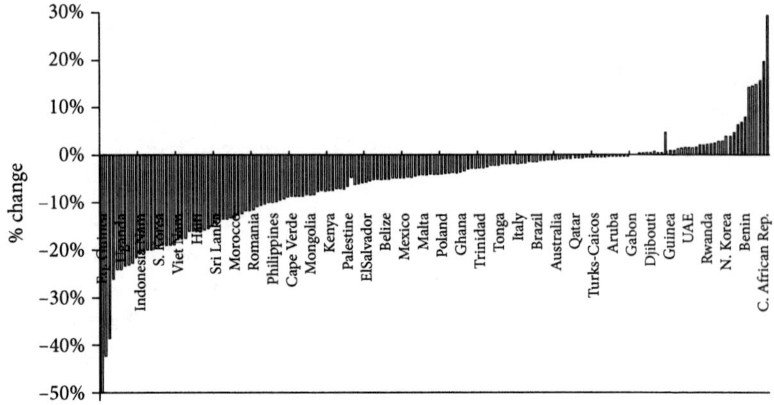

FIGURE 5.1 *Total percentage change in agricultural share: all countries, 1970–2008*
Source: Derived from UN National Accounts Database.

toward greater industrial progress. A handful have shifted by two or three percent, but the vast majority have changed enough to have altered their basic character. The strongest shifts also took place in the progressing group, with a decline of up to 54% in agricultural share within one country, Equatorial Guinea. This is a tiny nation, showing that this kind of drastic change is not dependent on economic scale. The second most extreme shift came from China, with a 42% decline in its agricultural sector. This marks nothing short of overturning the basic makeup of the world's largest country within a generation or two. In a nation reputed for slow and cautious long-term decisions, this is truly revolutionary.

Where are those economies that are shifting toward more agricultural concentration? Do they form a distinct geographic pattern? Figure 5.2 shows the geographic distribution of the world's 177 countries.

Those 137 countries with a declining share of agricultural output are shown as un-shaded territories on the map. They resist easy classification as they exhibit a diversity of geo-economic traits, as different as tropical Sri Lanka and temperate Canada, or polar Norway and Saharan Sudan. They are scattered through all regions, have vastly divergent physical, demographic, and economic scales, and range widely in their industrial structure. All of the world's economically powerful nations are in this group, growing away from agriculture.

Conversely, those 40 states with a rising share of agriculture are predominantly tropical, small, poor, islands, or oil exporters, while over a third of these are African. Many are struggling, poor, and declining

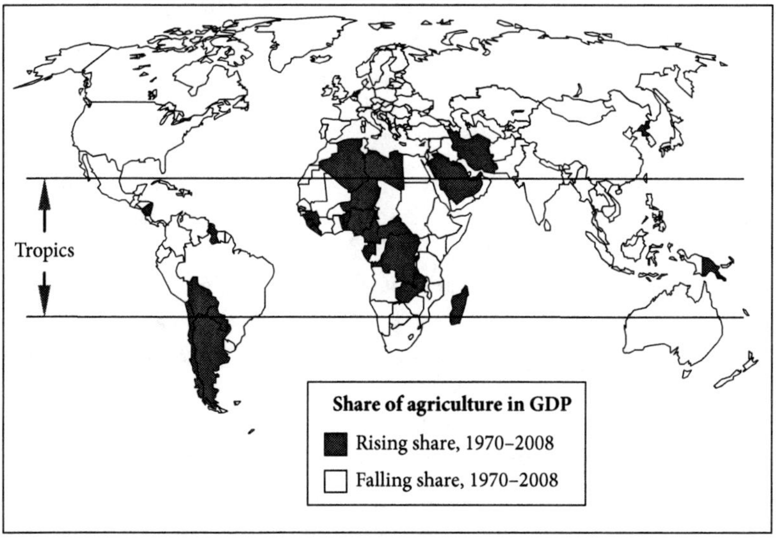

FIGURE 5.2 *Rising/falling agricultural share by country, 1970–2008*
Source: Derived from UN National Accounts Database.

economies such as Liberia, Djibouti, Haiti, and Somalia. Of this subgroup, Liberia and Haiti have seen a grave economic decline in which a growing share in agriculture accompanies a net decline in agricultural output, and the near decimation of their manufacturing sector. Their agricultural share is rising due to an overall economic decline as opposed to an improvement in agriculture. Djibouti and Somalia managed to maintain net growth in both sectors over the period, albeit at a rate low enough to be nearly negligible for a very low income economy (growth is about 1.8% per year, but GDP per capita is less than $2 per day).

A rising share in agriculture usually indicates a deteriorating economy and falling GDP per capita. Exceptions do exist. Rising agriculture can, in rare cases, be associated with an improving economy. This can happen, for instance, if an economy grows due to increased exports of oil or industrial goods that provide the capital to increase domestic agricultural investment. For example, from 1970 to 2008, Iran, Saudi Arabia, and Chile each showed an increase of about 700% in real agricultural output, and over 300% in manufacturing.

The large, stable, advanced countries do not exhibit rapid change. The Group of Eight Industrialized Countries, or G8, typically have an unchanging industrial structure in which agriculture has settled at a very

DOI: 10.1057/9781137340313

low level, and a GDP per capita that grows steadily but slowly. For example, the US lies very near the world average with an annual economic growth rate of about 2% over the past few decades. In contrast to the US, China's economy is neither stable nor advanced. Most economists consider its growth rate as too rapid to be stable, while the high share of agriculture in its structure (10%) is too high to classify it among the highly advanced. This is not to say that China has not made significant industrial progress, but rather that its high rate of change shows that it remains in transition. Although it is already well known that growth in advanced economies stabilizes at a low level, it is now becoming clear that industrial structure stabilizes with low agricultural share.

What does this mean for economic prediction? Can this shift of power toward China be quantified? Until now, it may have meant that China would eventually settle down from its rapid GDP growth rate of 8% to 10% per year to about 2%. The remaining question is how long this would take, but there is no reliable guide. The rate of industrial shift may bring some added insight. China's current transition resembles the past transitions of Sweden or the US, so it can be used to estimate the time required to reach the 2% level that indicates a likely stabilization. It must be stressed that this kind of projection does not produce highly reliable results, but may give a rough approximation. China's agricultural share has taken decades to fall from 50% to 10%, and its success in manufacturing suggests that it will continue falling. The coming years will likely differ for one key reason though, namely that recent growth led by the export of goods will slow down as world markets become increasingly saturated with those goods. Furthermore, as China's productivity rises, so will its prices. As a result, the country's transition to 2% agriculture should take decades more. And if other countries can be taken as a rough guide, its GDP per capita will continue to accelerate with its productivity which may in turn spawn a domestic consumer market. In a fully populated world this upward economic spiral will soon become constrained by energy and mineral resources, an unprecedented situation in human history.

Mapping the manufacturing shift

The majority of change in the world's manufacturing base clearly involves only three regions: North America, Europe, and Asia. Figure 5.3 captures this part of the global transition.

DOI: 10.1057/9781137340313

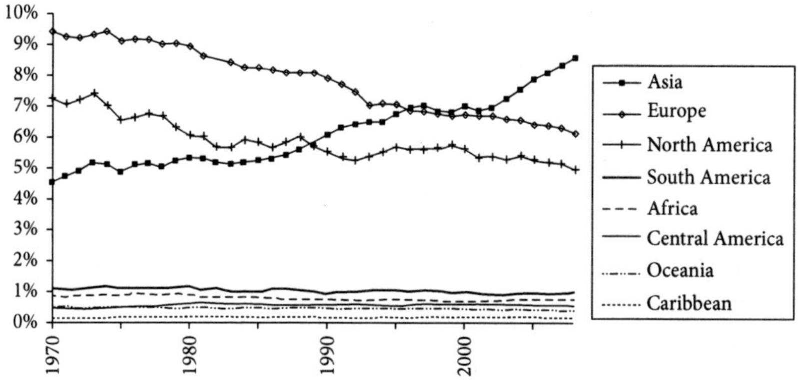

FIGURE 5.3 *Change in regional manufacturing as a share of world GDP, 1970–2008*

Source: Derived from UN National Accounts Database.

The decline in the share of world manufactured goods produced by the mature Western economies accompanies a rise in the share of world goods produced by Asia. Meanwhile, all other regions combined represent a small and mostly stagnant proportion of global production.

Under the UN regional classification, Asia is comprised of 51 countries. Among this group, Japan experienced the biggest post-war manufacturing expansion, rising from near total destruction to dwarfing China by more than ten times as of 1970. Its transition to the world's second ranking economy soon spurred Western journalists to dub Japan as an economic miracle, and even to predict that it would trigger the decline of the West.

Japan's manufacturing output also enjoyed continued growth along with its GDP until its pre-recession peak in 2007. In addition to a stable reputation for product excellence, Japan's economy had a unique trait of its own making. Its currency remained very strong even though interest rates were continually forced down by it central bank. This is, of course the opposite of the usual situation, but it was made possible by avoiding foreign debt. Japan's government was able to fill its borrowing needs mostly from its own citizens because of their unusually high savings. The strong yen eventually forced manufacturers to build plants to produce goods in their export market countries. The connection between savings, foreign debt, and exports will be discussed later in another context.

Regardless of all of these achievements and advantages, Japan's GDP growth has slowed to near stagnation since 1991, perhaps the clearest

DOI: 10.1057/9781137340313

example of the fading growth that all transitional economies experience. Its industrial shift had run its course with agriculture down to 1.3%, its GDP growth limited by an aging population and a resistance to immigration. Meanwhile, Western attention had already turned to China as a counterbalance to the Soviet Union's raw military power. Starting around 1980, a strategic Western economic engagement with China diverted some trade from Japan, possibly limiting its growth. Ultimately, economic power had stopped shifting solely from the West to Japan, and started shifting to the other Eastern centers as well. Partly to avoid handing excessive regional power to China, the US had also engaged in the economic build up of Taiwan and South Korea.

South Korea and India rank next in regional goods production, yet make up just a fraction of Japan's output. Meanwhile Japan's goods production remains at just one-half of China's. China has grown so much as to account for nearly half of Asia's real manufacturing output growth, and is now the main story anchoring the West–East shift in economic power. To illustrate the significance of China's manufacturing growth, think of it as having added more than another Japan within the span of just 30 years. From another perspective, the US share of world goods production fell from 27% to 21%, while China's rose from 1% to 12%.

Country shares and rankings tell one interesting part of the overall story. The other part is the size of the production pie. Has the global pie grown as China has grown, or has China taken part of the pie from other countries? China and the US each added about $1 trillion to the world's annual manufacturing total since 1970.[2] That global total grew from $3.2 trillion to $9.6 trillion in that period, meaning that China is likely adding to the global total instead of merely absorbing other countries' lost output. After all, the global total growth of $6.4 trillion was far greater than China's gain.

Identifying any losses in the manufacturing output of other countries would also help verify whether or not China gained at their expense. Did any country lose real output? Virtually none. Of the UN-listed countries that have been in existence since 1970, 96% have increased their real manufacturing output. The rare exceptions, those six that have lost real output, are mostly tiny and poor, or fraught with mismanagement and strife: Guyana, Haiti, Kiribati, Liberia, Sierra Leone, and Zimbabwe.

One final measure could help determine whether China's gain came from an existing base. That is, from a decline in growth elsewhere. For example, did the US manufacturing growth rate decline in recent

DOI: 10.1057/9781137340313

decades? In fact America's growth in manufacturing tells an unexpected story. First, it has never stopped growing, which may seem unlikely because it is seldom reported. Second, US annual manufacturing output not only grew, but did so at a robust rate, having doubled or added about 1$ trillion compared to 1970. Third, the sector's growth even picked up its pace starting around 1980, the year that China was granted Most Favored Nation trade status. From any perspective, China's rising factory output does not seem to have come at the expense of other countries.

Rank and decline

If manufacturing is power, then a look at the world's top four factory output rankings will show how the West–East power shift has played out since 1970.

China has entered the top tier as the clear new growth leader, so the popular perception is correct in that respect. But since the US and Japan have not surrendered the top two ranks, the power shift in manufacturing must have involved other countries more directly.[3] Germany is also touted as a rising global powerhouse but it declined from third to fourth place. In 1970, Germany's third place output was virtually tied with Japan's for second place, but far ahead of the UK. In a relative reversal of fortune, Germany had fallen far behind China by 2008. Although it remained Europe's so-called workshop by a wide margin, Germany's relative output skidded to half the size of China's and Japan's. It still grew, but the Eastern powers grew much more quickly. Within the major Euro region nations, the UK, Italy, and France, with achieved modest manufacturing growth but South Korea is closing the gap with these countries very rapidly. With China and South Korea making a strong entry into the top ten, the real West–East power shift in this sector will be from Europe and the US to East Asia.

TABLE 5.1 *Factory output rankings, 1970 and 2008*

	1970	2008
1	US	US
2	Japan	Japan
3	Germany	China
4	UK	Germany

DOI: 10.1057/9781137340313

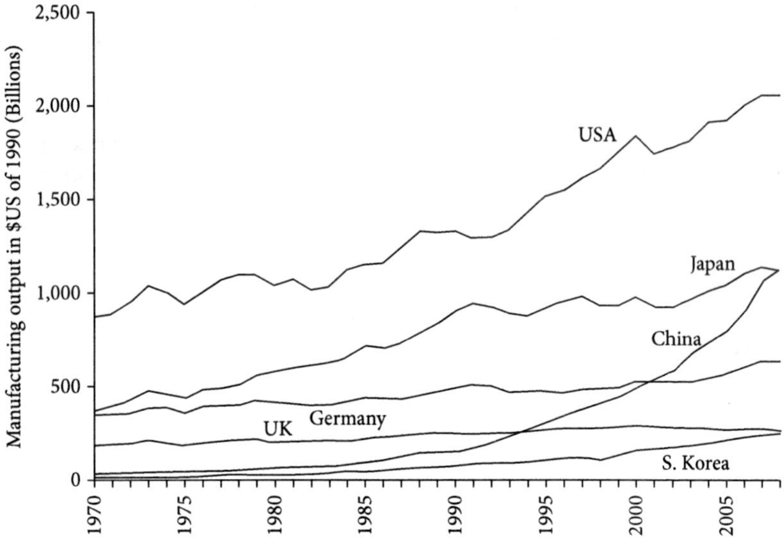

FIGURE 5.4 *Change in rank of manufacturing countries by size, 1970–2008*
Source: Derived from UN National Accounts Database.

The chart in Figure 5.4 summarizes the manufacturing output trends of the world's top producers.

Most importantly, the chart shows that the US and China each added about $1 trillion to their annual manufacturing output. Although US manufacturing growth rarely gets attention, this increase represented more than a doubling of US output. China's increase was 38-fold, which can also be realistically seen as a rise from zero, and an overtaking of every Euro zone country. China's size now approaches that of the combined Euro zone countries.

Does the BRIC group matter in terms of its manufacturing output? The members' GDP rankings among the world's nations are mixed as Brazil has not moved, and Russia has fallen behind while India and China have advanced. The entire group's output is about 10% less than the US' alone. Although India has grown very quickly, to about ten times its 1970 size, it remains at about one-tenth of the US and about one fifth of China. Seen in yet another way, India's manufacturing sector is about the same size as Canada's and does not rank in the top ten. Without China, the group's output is about the same as Germany's, suggesting that none of the emerging economies is likely to overtake today's industrial giants in the near term.

DOI: 10.1057/9781137340313

China became the largest Asian manufacturer in 2011. At the onset of the 2008 global financial crisis, manufacturing output in China and Japan was about equal, but the following three years brought a major inversion. As the established economies slowed for four years, China grew until it exceeded Japan by about 80% and assumed the world's second rank. This recession may cement China's position, or a global recovery may produce some rise and fall in the rankings. In either case it seems clear that China will always be a central element in the world economy. Interest in China as an industrial economy builds upon a very long history, and it has always figured prominently in studies of East–West power. Western trade connections with China date back to the 16th century, with deep impacts from Britain in the 19th century. Part of that interest hinged on the gigantic scale of the country, both in landmass and in population. It was simply too big to be ignored for the military danger it could impose, and the industrial potential it could present.

Understanding China's economy would almost always be a problem though. Until 1970 there was no global organization of states and no consolidated data reporting such as today's UN or World Bank. Even as late as the 1950s (and in some ways, even today) assessing China's industrial capability involved rough estimates, guesswork, and loose assumptions based on trusted personal contacts. Who was to know how much silk cloth was made each year, or how many oxen were employed in farming? American economist Simon Kuznets (inventor of the concept of GDP) compiled exactly this kind of information and worried a great deal about how to compare economies that stood at different stages of development. And how different they were. In the 1950s, the US led the industrial world while China starved. Detroit churned out millions of cars each year while China produced a handful of farm tractors. Today China's annual auto production exceeds America's. Chongqing has grown to rival almost any modern city on earth with a metropolitan population of about 35 million, while Detroit struggles to curtail the fleeing of its last 1 million core citizens. There is no doubt that some measure of industrial power has shifted from West to East. The story, however, is not as straightforward as that.

The decline of the West may turn out to be exaggerated. In its own right, it is still a major manufacturing force as measured by growing output as well as technological advancement, so the idea of decline is currently incorrect. Even though the West has never stopped growing, China's rise in manufacturing has been interpreted as signaling the

DOI: 10.1057/9781137340313

Western decline in relative terms. While relative size is indeed changing, the rise of China can also be viewed in another light; it demonstrates the enormous economic power contributed by the West toward the resurrection of Eastern industrial capacity. China's industrial potential has essentially lain dormant for decades compared to the rest of the world during the Cold War, but received an enormous boost from US trade policy recognition. This was a strategic political move by the US that turned out to have economic consequences of historic proportions.

After watching the Soviet Union grow its industrial capacity from near zero to superpower status in about thirty years, Western nations realized that a major new possibility had opened in the world. By brute force, Joseph Stalin shifted Russia from farm to factory to satellites in space; it was clear that China, perhaps under Mao Zedong, could do the same. More crucially, if a newly industrialized China was to ally itself with Russia, it would pose a serious threat to global stability. After all, Russia had gained several Asian client states after the Second World War, and followed with virtual takeovers in Eastern Europe. Poland, Hungary, Czechoslovakia, and several others fell under Soviet Russia's influence or control. Some measure of political power had shifted eastward to Russia, but it remained to be seen whether economic power would follow. So one choice was to ignore China, and risk that it would take over several smaller Asian nations, perhaps Burma, Laos, Cambodia, Thailand, Korea, or Vietnam. This path also carried the larger risk that Russia and China would take the standard geo-political maneuver of allying themselves as a counterbalance to Western military power. The other choice was to engage China politically and economically, by opening diplomatic channels and accepting its export products. Richard Nixon chose the second path in 1972, and set in motion a West-East power shift that differed fundamentally from the previous internal shift by Russia. Nixon's was a deliberate act to forestall a new military enemy by creating a new economic partner. China attained Most Favored Nation status in trade within just eight years.

In Soviet Russia's case, industrial growth resulted from a 'big push' internally, a strategy that came directly from Premier Joseph Stalin. The push lasted from 1922 through the early 1950s, with major developments in infrastructure, manufacturing, and military technology, all at the cost human life and dreadful results in agriculture. But underlying the gigantic industrial advance was a total antagonism with the West. China's case was almost completely different in character. China's industrial shift was

DOI: 10.1057/9781137340313

fully coupled with the economic policies of the United States. Whereas US trade with Russia was almost negligible, trade with China is a policy cornerstone. Even though serious problems accompany US–China trade, dialogue and progress are ongoing. Some accuse China of having based its gains in recent decades on a few key ill-gotten external elements, including Western technology, Western export markets, and Western cooperation. This view ignores the crucial fact that the opening was made by the West, and that the original reason for that opening was as an intentional counterbalance to the USSR. Even if currency manipulation and weak patent protection create trade irritations today, America in 1970 fully understood that it was about to offer an economic relationship to a pre-capitalist China. America stood to gain the most important ally available in the world, and possible trade irritants simply paled in comparison.

Because of its size and rapidly increasing industrial capacity, China rightly gets much of the world's attention while setting the public mood surrounding trade issues. Although its scale does merit a great deal of focus, the singularity of the world's obsession may be overdone. For example, as a share of the 2011 US trade deficit, China accounts for 41%. As large as it is, this deficit remains only 1.2% of US GDP. The US also manufactured $2.3 trillion worth of goods in 2011, 28% more output than China (as measured in constant $US of 2005). Although the popular perception may be that the US has given up on manufacturing, and seems to prefer importing such goods, this is simply not the case. Total manufacturing output has increased continually for nearly all of the past century, and has done so with ever increasing productivity. The rise of China is not the fall of the US.

A recent report by the Federal Reserve Bank of San Francisco provides some detailed insight into the extent of China's exports to the US. The report shows that "Chinese goods account for 2.7% of U.S. PCE [personal consumption expenditure], about one-quarter of the 11.5% foreign share."[4] It also reports that imports to the US totaled 16% of America's GDP whereas China's contribution to that share was only 2.5%. There is no doubt that the US has a serious trade imbalance, but its trade with China receives more than its fair share of the blame. That is likely due to the kinds of goods that originate in China. These goods are highly visible to vast numbers of American consumers: household goods, furniture, shoes, and clothing. The story also takes a somewhat unexpected turn in the services sector. Instead of a trade deficit, the US produced a surplus with every major trading partner, including China in 2011. The yearly balance in services was a surplus of $194 billion, $15 billion of which came from China.

The service revolution

Over the past century, the economic mindset was that progress involved increasing a nation's manufacturing capacity. Advances in technology would make firms more productive and free up labor to engage in new activities, such as inventing yet more machines, improving business processes, or training workers. This would also spin off new products which could be exported to other countries and earn more income in foreign markets. As a result, the production of goods could just go on expanding and form the basis of growth in the Western world. The standard of living was to improve in two ways simultaneously: foreign countries would enjoy new and improved products, while the industrial economies enjoyed rising income.

What was the anticipated effect on industrial structure? In the post-Second World War decades, the possibilities seemed limited. In a typical example of the risks associated with prediction, economists deemed the world's industrial structure to be largely settled. Manufacturing could continue its gradual growth while agriculture slowly faded in importance. In effect, the first two sectors could trade their shares, but services were predicted to have reached a sort of upper limit. A key constraint on the service sector was thought to be that services could not be traded or exported, and in the end, this was the key mistake that supported a faulty prediction.

Decades ago, students learned that services differed from goods in one important way; they were not tradable. Their lessons often featured silly examples such as the impossibility of exporting haircuts. This could happen only if a barber crossed the ocean on a ship to cut someone's hair, but of course the cost made it economically impossible. The fun of ridiculing this specific possibility masked the broad reality for a few years, but two emerging technologies were poised to change the entire world. In the 1950s, computers and jet powered aircraft had very little practical application on a large scale. Jets had recently shifted from military uses, such as bombers, to civilian flights, although they were expensive to fly and prone to disastrous failures: their frames cracked and their

DOI: 10.1057/9781137340313

engines erupted into flames. Computers also had a long road to becoming a mass-produced commodity. Early versions of simple machines in the 1950s weighed hundreds of pounds and filled about the same space as a domestic living room. In retrospect, there was little to suggest that airplanes and computers were about to transform the structure of the world economy.

By the late 1960s, jet airplane technology advanced to the point where flying became the safest way to transport people. Highly purified aluminum drastically reduced airframe failures, while engines, tires, and every other component were refined to the threshold of being foolproof. The Boeing 747 jumbo-jet was launched in 1967, and in turn launched a miniature industrial revolution. Here was the first step in enabling the export of 'haircuts'.

Mass recreational air travel became an entirely new category in the services sector, spawning more international hotel operators, airports, ground transport firms, and a wide array of related services. Affordable business-related air travel now meant that professionals could regularly consult with international clients face-to-face on medical research, engineering projects, financial transactions, and legal matters. The range of services that could practically be sourced overseas grew ever wider. Today, a somewhat weird and wonderful example links the farm to the jet age. The so-called flying winemakers are those highly sought superstars who jet between hemispheres consulting to vineyards willing to purchase their services. Their only stock in trade is modern knowledge of an ancient art, and they export it at the speed of sound to clients around the world. The global industrial shift had brought agriculture, manufacturing, and services together in an entirely new way.

Earlier precedents in global trade are long forgotten, but had similar impacts. Refrigeration and steam ships may not seem highly relevant today, but when they were first combined, distant economies collided. In one example, Argentina collided with France in a 19th-century episode of "exporting land". The production of beef requires a great deal of open land, so wherever land is relatively scarce, such as France, the cost of beef is high. The new technology enabled Argentine beef producers to ship their frozen product to Europe. This meant that a land-rich region in the southern hemisphere could now send its products to a land-poor region in the North. In effect, Argentina's abundant land was exported to France resulting in cheaper beef. French consumers welcomed the lower cost product but French farmers were enraged. Economists are fond of

DOI: 10.1057/9781137340313

saying that the solution for globalization is globalization. Beef imports caused such a stir that French tariffs were soon raised in order to choke off the new supply.

Over a century later, many Americans emulate the French, complaining to their government representatives that cheap goods from China have caused higher unemployment in US manufacturing. In effect, three industrial shifts coincided and changed the world. In China, rural workers moved by their millions from small farms to urban factories. Call it the great shift forward. Next, business firms from every corner of the world clamored to open manufacturing plants, injecting enormous financial and physical capital into the Chinese economy. Finally, China began exporting vast quantities of low cost goods to the world's markets. Western consumers saved money on their routine goods, and low prices helped to stanch inflation. In effect, a huge new labor force has been exported from China to the rest of the world using another transport innovation, the standardized shipping container. The news media refer to this as the "hollowing out" of the American middle class, and often call for the same response as the French farmers previously demanded.

Import tariffs might seem like a straightforward remedy for foreign competition, but economists have a stock reply for this too: beware the unintended consequences. The standard analysis shows that tariffs bring higher prices to consumers, and that domestic firms produce lower quality goods without the pressure from competitors. But in China's case, retaliation with tariffs would be even worse for the US. If import prices were to rise by a moderate amount, consumers would be forced to pay more for Chinese goods but they would still avoid high priced US goods. Consumers would be worse off, with the situation for US producers unchanged. Instead, to provoke a switch to buying American, tariffs would be need to be very high and prices would rise steeply, at which point the consumer demand gets choked off. Again, US firms would not gain new customers, and they would not hire more workers.

The export of services, once seen as virtually impossible, may turn out to be the greatest imminent threat to the Western economies. Exporting costly specialized services helped to provide the funds needed to purchase imported goods. But this also helped to stoke the tide of cheap goods from low wage foreign manufacturers. The new risk is that low wage foreign economies can now sell cheap services to the West. And just as 19th-century France responded to the refrigerated shipping of beef from Argentina with cries for protection from foreign competitors,

DOI: 10.1057/9781137340313

a significant amount of public sentiment in Western economies now rails against the ongoing power shift. Farmers cried for protection, manufacturers followed, and services firms will soon follow too.

A taste of the new world economy

In one of the clearest examples of how unclear the situation is, China has become a world leader in genetic sequencing by purchasing the needed equipment from the United States. San Diego-based Illumina Inc. manufactures the sequencing machines that major universities and governments use in their medical and industrial research. The medical community envisions that such technology will lead to so-called personalized medicine, in which the treatment of a disease is tailored to suit the individual patient instead of the general population. Drug formulation and dosage becomes highly specific, and the chance of adverse reactions is reduced, which in itself signals a major advance in medical science. So contrary to the widespread perception of the demise of US manufacturing, at least part of the transformation within the sector has been to move from common technology to the highest level of advancement. And contrary to the doomsday scenario that focuses exclusively on cheap Chinese imports, this firm has demonstrated the fact that China can also become an export market. In 2010 the firm announced its single largest order, 128 units, sold to a scientific services company in China[5] (at about $700,000 per unit). In this single transaction, a leading US high science manufacturing firm supplies the capital goods that enable a Chinese firm to become a leader in high science services. Likewise, Canadian manufacturing firms supply China with high technology equipment such as flight training simulators, regional jet aircraft, high speed trains, and satellite imaging, all to the services sector. The East–West power shift runs in both directions, and this too is a part of the new world economy.

Notes

1 See Appendix C for a complete list. The field is limited to those 177 countries listed by the UN that have been in existence since 1970. The main exclusions in the figure are the former Soviet states that (re)gained independence after the 1990 collapse, and, the only state with no agriculture, Monaco.

2 The figures in this section refer to constant $US of 1990 quoted in the UN National Accounts online database.

3 Japan is on the verge of being overtaken by China in 2008. Recent reports state that China has surpassed both Japan and the US in factory output in 2011. While there is no doubt that China has increased its output faster than most countries, and its rank is rising, there remains some controversy about surpassing the US. Reports making that claim are based on current prices (not adjusted for inflation). Using current prices, China's factory output exceeds the US by about 20% in 2011. Using constant prices of the year 2005, the US still exceeds China's factory output by 28%. All figures from the UN.

4 *FRBSF Economic Newsletter 2011–25*, Federal Reserve Bank of San Francisco (FRBSF), August 8, 2011.

5 http://investor.illumina.com/phoenix.zhtml?c=121127&p=irol-newsArticle_print&ID=1374343&highlight=

DOI: 10.1057/9781137340313

6

Power Shift: North and South

Abstract: *The Caribbean shows that rapid industrial shifts into services can produce rapid economic growth. Africa and Latin America remain a perennial concern with respect to development. They also remain a very small and virtually unchanging proportion of the world economy. Brazil's outstanding growth is the main exception to South American stagnation, even though its agricultural sector has surged in size and share. Pockets of growth are evident in Africa although new mining and oil industries bring little technological advancement. The majority of small, poor, stagnant economies still focus on agriculture, and they lie in Central Africa.*

Climate issues are largely ignored by economists, although they are demonstrably important.

Key words: North–South issues; North–South trade

Atikian, Joe. *Industrial Shift: The Structure of the New World Economy*, New York: Palgrave Macmillan, 2013.
DOI: 10.1057/9781137340313.

The missing shift

The North–South Institute bills itself as Canada's leading think-tank on global development issues and poverty reduction, conducting research on policies aimed at improving development in poor countries.[1] The institute provides no indication as to why the name refers to North and South, but one might presume that it involves a northern hemisphere viewpoint of the poor South, an idea that may spark some indignation from Uruguay, Australia, New Zealand, Aruba, Singapore, Brunei, and Brazil, even allowing the best of intentions.

While it is true that most poor countries lie to the south of most wealthy countries, it seems to be a mere accident of geography because no one has yet assembled a credible explanation for this plain fact of economic life. More land exists in the temperate climate regions of the northern hemisphere than there is in the South, but there remains a great deal of conflict as to why this should matter to a nation's wealth.

Educated views on national development conflict as widely as they do on the relevance of geography. Columbia University professor Jeffrey Sachs works on development, foreign aid, and the direct involvement of wealthy nations. Former World Bank economist William Easterly rejects this approach as being blind to foreign aid's development failures across the decades around the world. Zambian economist Dambisa Moyo agrees with Easterly and supports private sector approaches, while musician/philanthropist Bob Geldof once called for aid donors to "give me the money" so that he could give it to the poor.

The problems on the ground are real and serious, as multitudes suffered for years through various marginally successful development schemes. The perspective from industrial structure may offer a new way to compare northern and southern regions. These two categories are admittedly ambiguous, at least partly because poor nations exist in both hemispheres. They are, however, suitable in the sense that after decades of usage they are now widely accepted. And perhaps after the past few decades of varying development results, some southern regions can shed light on the problem for other southern regions.

In stark contrast to the dynamism seen in the East–West power shift, very little has shifted in the North–South direction. Asia's rise in manufacturing clearly dominates the global scene, while most of the southern regions stagnate at a low level. For example, Africa's share of agriculture in GDP has risen by a modest 3% over four decades, and this is not an

DOI: 10.1057/9781137340313

TABLE 6.1 *UN Regional Groupings*[2]

South Asia	Southeast Asia	East Asia
Afghanistan	Brunei	China
Bangladesh	Cambodia	Hong Kong
Bhutan	Indonesia	Macau
India	Laos	DPRK (N. Korea)
Iran	Malaysia	Japan
Maldives	Myanmar	Mongolia
Nepal	Philippines	S. Korea
Pakistan	Singapore	
Sri Lanka	Thailand	
	Timor-Leste	
	Viet Nam	

economically positive sign in most areas. Entire continents with millions of workers in a hyper-connected world of air travel and internet have barely made a step toward shifting world economic power from North to South.

In a significant exception, a closer look at two sub-regions gives a more optimistic view. These are South Asia and Southeast Asia (see Table 6.1). India dominates the first sub-region, while a group of important smaller countries comprise the second.

The island way

Three southern regions have shown major industrial progress in substantially different ways. South Asia and Southeast Asia have seen a classic shift from agriculture to manufacturing. In contrast, the Caribbean has taken an unusual path and to some extent transitioned more directly from agriculture to services.

The Caribbean presents a unique group of independent nations on a unique industrial path. Many of these small island nations started as colonies, endured a great deal of hardship and conflict before gaining national independence, and finally built their industrial base as import–export economies. As with most economies worldwide, they began with

DOI: 10.1057/9781137340313

a focus on agriculture and mining before slowly shifting toward manufacturing. But as the jet age spawned a new services-based sub-sector, tourism, many of the islands shifted more intensively in this direction, also finding new opportunities in international finance. Hotels, banking, and insurance helped several Caribbean countries to bypass manufacturing and generate very high income. The regional share of GDP generated by agriculture fell by over one-half, to a mere 3.2%.

Although the overall scale of the Caribbean is small, it provides an example and a path for a North–South economic power shift. Did some of this shift also stem from the alignment of national independence dates with the jet age dawn? It did for many of these nations, as this timing may have played a role in accelerating their services sector growth.[3] In retrospect, the combination of aircraft technology and the Caribbean's location in a pleasant climatic zone has dealt the region a fortunate hand. The new aircraft delivered new tourists to the newly independent countries, and contributed to their industrial shift. If, instead, these countries had followed the traditional shift from agriculture to focus on developing a significant manufacturing base, their small size would likely have rendered them uncompetitive. In a similar setting, the so-called Malay Archipelago is home to nations such as Indonesia and Malaysia, with enormous populations that have shifted to manufacturing. Since scale has a crucial influence on manufacturing cost, the Caribbean may have stood little competitive chance.

The four largest islands in the broader region belong to a sub-group the Greater Antilles. These include Jamaica, Cuba, Puerto Rico, and Hispaniola, which is shared by Haiti and Dominican Republic. Puerto Rico's industrial development and income level far exceeds the others owing to its full economic integration with the US. Among the remaining four nations, three could be expected to bear economic disadvantages such as political instability in Haiti, recent revolution and dictatorship in DR, and a fifty-year trade embargo by the US against Cuba. In contrast, Jamaica has had a continuous democracy, is free to trade with every nation in the world, has an English language advantage over all of its neighbors in US markets, and has even retained its membership in the British Commonwealth since its 1962 independence. And yet with all of these advantages, its economic growth per capita steadily deteriorated for over half a century. Jamaica's economy has virtually stalled over this period whereas even Cuba has generated modest per capita growth. One hint at the reason comes from industrial structure.

DOI: 10.1057/9781137340313

Over the course of four decades, Jamaica's share of GDP generated by farming has stagnated. It started the period at 6.4% and ended with 6.1%; it is as though nothing has happened to its industrial base. In contrast, DR's agricultural share has fallen from 21% to 7%, and Cuba's has fallen from 13% to 6%. And, even as poor and aid-reliant as Haiti remains, its farm share moved from 44% to 28%. All of the regions with declining agricultural share experience per capita GDP growth, and the Caribbean is no exception. The relationship continues to hold true for regions around the world, as well as the majority of individual country examples.

From crisis to stagnation

Africa has increased the share of farming in its regional economy, which runs generally against the wealth building trends in the rest of the world. Although it showed promising signs of shifting away from agriculture during the early 1970s the brief trend reversed itself and ended up unchanged. Meanwhile the continent's per capita income level followed the world's historical trend and increased only very slightly, lagging the growth seen in most other regions.

This combination of stagnant industrial structure and income nearly mirrors the situation in South America, except that Africa's case is worse due its generally lower development level. Low productivity in agriculture has a firm grip on Africa's economies. South Asia and Southeast Asia started the same period in much worse industrial condition. Their agricultural levels in GDP were much higher and income was much lower. How much worse off were they? Double, and half. That is, the farming share in GDP doubled Africa's share, while Asian income stood at half of Africa's. Both factors have since improved to the point where South Asia's income now matches African income, and Southeast Asia's income doubles it.

Figure 6.1 gives an overview of these southern regions strictly from the perspective of shifts in their industrial structure. North America is included as a comparator, and to emphasize the strong relationship between declining farm share and rising wealth. It has by far the lowest farm share, and the highest wealth.

Intersecting lines on the chart point to a change of rank among the regions. While Africa's horizontal line shows its stagnant structure, the

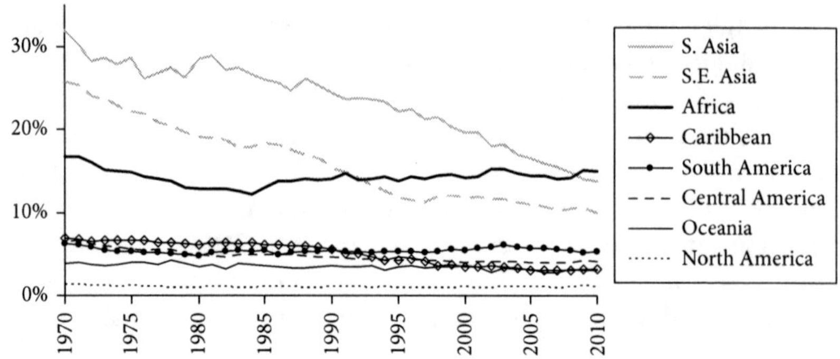

FIGURE 6.1 *Trend in agricultural share by region, 1970–2008*
Source: Derived from UN National Accounts Database.

sloping line for each Asian region indicates a rapid shift away from agriculture. Even though South Asia's annual agricultural output doubled, manufacturing increased by even more, and so did its economic wealth.

Africa has been left behind by regions that were recently just as poor or worse off. It may seem that Asia did not in fact start off in a worse position than Africa, but in 1970 Southeast Asia buckled under war. Laos was carpet bombed by American forces trying to prevent communist military incursions. Cambodia followed its bombing episodes with a self destructive campaign intent on returning the country to its distant agrarian roots. Viet Nam split in a de facto civil war waged by foreign forces. The region's largest country, Indonesia, labored under successive dictatorships, while Burma soon turned into one. To think that Asia was at an economic advantage over Africa perhaps looks too lightly at the situation on the ground. Even so, the economic power shift toward the East has no counterpart in the South.

Africa's share of world manufacturing has remained mostly stable for decades, although a major trend hidden in this data is clearly visible on the ground: that is, a major reversion toward the hardest hitting extractive industries, crude oil, and large-scale mining. And these are not necessarily hard hitting in an economically beneficial way, but rather in a socially destructive way. Governments can easily hoard the proceeds of resource exports while the ecological impacts and immediate pollution can go unchecked.

How big is Africa's mining sector compared to its manufacturing? Vastly larger than any other region in the world. The world's overall average mining sector comprises about 30% of the manufacturing sector's output. Among the world's regions, mining ranges from 20% to 80% of the manufacturing sector. The only region beyond this range is Africa, and its mining sector is 202% larger than manufacturing.

This shows the extent of Africa's dependence on simple resource extraction, which, furthermore, differs in character from mining in the more wealthy regions and countries. For instance, in Canada, Australia, and the US, mining and oil extraction is heavily dependent on various advanced technologies, including electromagnetic exploration, steam assisted pumping, and hydraulic fracturing. These technologies were not imported from yet further regions, but invented and developed from within. In contrast, Africa's mining sector is heavily reliant on basic foreign technology, whose usage results in some learning but very little technological developmental.

~ ~ ~ ~

Figure 6.1 seems to show South America as a region with a completely stagnant industrial structure. Its agricultural share in the economy hovered around 7% for decades, more advanced than the world's poorest regions, but making no progress toward gaining the economic power that accompanies a strong shift toward the services sector. José De Gregorio, former Vice-President of Chile's central bank, offered a dismal synopsis of an important regional economic trend: "Latin America has been dominated by growth expansions that, more often than not, have ended in crises and protracted periods of stagnation".[4]

Much of the continent has been beset by a revolving door of political leaders, forcing their economies into swings between farm strikes and capital flight. Dictators have given way to democrats, only to return and collapse again. This political instability has bred economic crises, just as coup d'états follow questionable elections. And stable periods stemming from these strong-man presidents tend only to achieve economic redistribution instead of technologically based growth. One important exception to this pattern is also the region's largest country, Brazil.

Figure 6.2 reflects some of the economic effects of South America's political patterns.

Brazil's chart of industrial structure shows a series of reversions in agricultural share, decade by decade. The country made major progress in the 1970s as the sector fell from 13% to 9% of GDP. This shift appears

DOI: 10.1057/9781137340313

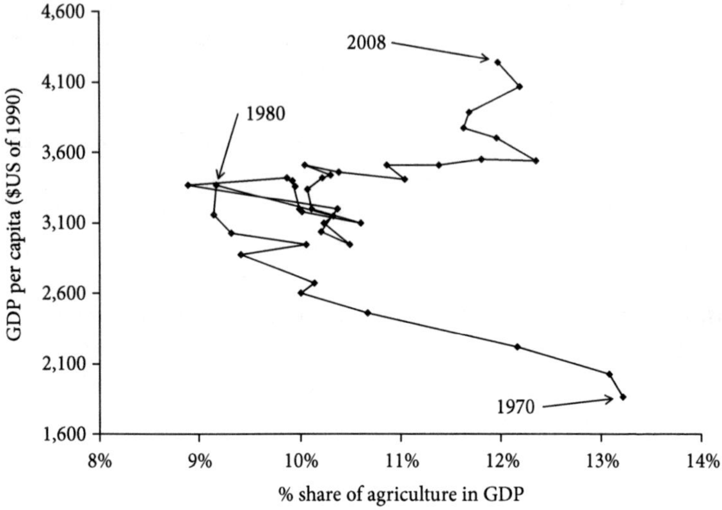

FIGURE 6.2 *Brazil agricultural share vs. GDP per capita, 1970–2008*
Source: Derived from UN National Accounts Database.

to be the result of huge financial capital infusions from several wealthy northern economies. Real farm output still grew by 60% in the decade, but the new capital flowed to investments in power production, roads, and various manufacturing firms including steel and autos; the factory sector simply grew faster than the farm sector. As the farm sector fell in relative size, national income increased by about 80%. Such a significant rise in the income of such a large country represents one of the first cases of a North–South power shift. A major world economy could emerge.

Brazil's smooth transition in 1970s was followed by a global interest rate crisis in the 1980s. On the chart, the crisis shows up as a rising agricultural share as investment in infrastructure and manufacturing declined. A decline in GDP per capita followed the initial interest rate crisis, and income remained stagnant for twenty-three years. This stagnation through the 1980s and 1990s doesn't usually show up in annual GDP figures. GDP almost always rises, at least in part due to inflation. But population also rises, and if it grows faster than GDP, then real wealth erodes. Brazil's progressive shift has restarted in the past decade, with a new leg of rising income on the chart. From 1970 to 2008, Brazil's farm sector share declined by a marginal amount, 1.2%, but its economy has advanced in technology and quadrupled in size, with a doubling of income per capita. This country is already a leader in the new South.

As of today, Brazil has proven to be a major success in developing a modern and diversified industrial economy at a large scale. It also stands alone not just as a southern power, but as a tropical economy. So, has Brazil shown that tropical location has no bearing on economies? Not quite. In one respect Brazil resembles Canada; its industrial development and population has clustered along a narrow strip of its territory, the furthest away from its most extreme climate regions. In Canada's case, this means concentrated building along its southern border with the US, as if to turn its back to the northern cold. In Brazil's case, industrial development is concentrated along the Atlantic shore and tends toward the more temperate South. It's not that industry has made tropical climate irrelevant, but that it deals with this geographic fact by choosing its geographic location as practically as possible.

~ ~ ~ ~

Much has been made of Africa's recent growth. GDP has risen threefold since 1970, and per capita income has increased although to quite a lesser extent. These scant statistics offer some optimism but they also distract from the full picture in the global context. Although Africa's GDP growth kept pace with the world average, it severely lagged Asia and its southern sub-regions that have taken up manufacturing. Worse yet, considering its booming population growth, Africa gained less than nothing in the share of world GDP. So not only has a North–South shift in economic power eluded Africa, the continent has fallen behind (see Table 6.2).

TABLE 6.2 *Focus on Africa's GDP and population growth*

	1970	2010
Population	366 million	1030 million
GDP	$0.336 trillion	$1.2 trillion
GDP per capita	$918	$1165
Share of world pop.	10%	15%
Share of world GDP	2.2%	2.4%

DOI: 10.1057/9781137340313

In order for Africa to have maintained its 1970 share of the world's $51 trillion GDP relative to its population, Africa would have needed 3.3% of world GDP in 2010 (instead of its actual 2.4%).

(All figures in this frame are from 2010, and GDP is in constant $US of 2005).

Source: Derived from UN National Accounts Database.

Geography has not helped Africa. The continent is physically disconnected from neighboring regions by sea, as the Mediterranean forms a barrier to Europe and the Indian Ocean separates it from Asia. The nearest region is the relatively poor and unstable Middle East, so good economic opportunities based on proximity do not exist. Nonetheless, geographic isolation never completely prevents economic progress; at worst it can hamper it to a limited extent. For instance, New Zealand lays at the most remote possible point from all major economic centers, operates with the disadvantage small scale, and yet manages a moderately advanced economy. The same holds for Iceland.

Resource extraction in Africa has accelerated as China expands its global activity in this sector, but as often occurs, the earnings can be instantly re-exported by the authorities. The Democratic Republic of Congo (DRC) suffers the ongoing effects of resource-related impacts combined with military and other conflict. How does this affect the progress of its industrial structure and wealth? A grave social distortion restricts the economic benefits of resource development to a narrow swath of society, leaving the majority with poverty and low human capability.

The DRC doubled its mining and manufacturing output from 1970 to 1990, before falling by two-thirds. This extreme instability characterizes the whole disastrous economy, but further details loom large. The historic situation for most countries usually entails mining as a small sub-sector of manufacturing. In the DRC, mining has taken over as the majority of the sector since the early 1990s, more than doubling its historic levels. Agriculture rose to a very high share of GDP (28% to 43%), indicating a possible collapse, as the country's GDP fell by a disastrous 40% before partly recovering. This kind of transformation in industrial structure typifies many African national economies, and its effects spill into the wider region.

Gabon has also had a mining-based resource boom, but centered on crude oil instead of mining. In a reversal of sorts, this economy

DOI: 10.1057/9781137340313

has demonstrated that a resource focus can bring a somewhat more broadly based prosperity. Its structure resembles some of the world's more advanced economies in that agricultural share stayed in a narrow range of 3% to 4% over 40 years. In a highly unusual departure though, its manufacturing sub-sector remained the same small size as agriculture. Gabon's mining (oil extraction) sector is also at an extraordinarily high share, about 50% to 60% of GDP through the decades. But instead of the expected disastrous result, came more of a modest success visible in two measures, the poverty headcount percentage, and life expectancy. Compared to DRC, Gabon's poverty headcount is less than half, at 33%, while its life expectancy is 63 years compared to DRC's 48 years.[5] While Gabon provides an example of Africa's capabilities and progress, the overall situation remains that the continent's rank in the world continues to decline, even if gradually. Economic power continues shifting from South to North in one of the world's largest poor regions.

The economic climate

One of the most obvious aspects of North–South studies consistently receives the least attention: climate. The field of economics almost completely ignores it. It's not that climate has gone unrecognized, rather that it is most often casually dismissed or lumped in with a larger group of physical features collectively referred to as geography. A perennial quarrel in economics revolves around the opposition between geography and institutions. Which one drives economic growth and societal development? Top experts fall on either side, but climate is almost unanimously seen as unimportant and even discredited. A stronger case needs to be made.

Every atlas shows that wealth is distributed according to climate around the world, but the same pattern goes unnoticed within each continent and within any given country. This seems impossible to those who believe that institutions are the major development factors: legal systems, political structure, shared language, customs, religion, and national history. The evidence is plain though. The hottest and coldest regions in the continental USA are in the central core, from the mouth of the Mississippi, north to the Canadian border. Think of Louisiana summers and North Dakota winters. These also tend to be the lowest income

DOI: 10.1057/9781137340313

regions regardless of the fact that they share all of their institutions with the rest of the country. There are no colonies, disparities in government corruption, or political structure to explain the gap in wealth relative to the far more prosperous and temperate climate states on the east and west coasts.

The same climate–wealth pattern holds within rich countries, poor countries, or developing countries, whether they are small, large, isolated, or surrounded. But doesn't Brazil have a vastly improving economy that refutes this? Not in its hottest tropical areas. All of Brazil's economic development lies on its Atlantic coast where the tropical heat is moderated by the enormous cooling capacity of the ocean. Don't Russia, China, and India have vastly improving economies that indicate flourishing development in extreme climates? Not in their coldest and hottest regions. The same pattern holds in small, dual climate countries such as Bolivia. Isn't Australia a hot, wealthy country? Not in its tropical or hot desert areas. Australia's wealth lies in a concentrated strip in the temperate South, which incidentally makes it one of the most remote trading regions in the world and helps refute the idea of high transport costs as a cause of severe poverty.

One special kind of climate outcome occurs almost exclusively in small countries. Normal climate events can devastate or destroy small economies. For example, a single hurricane can cripple a Caribbean island nation to the extent that it must rely on foreign assistance for its recovery. The same storm can cause far greater economic damage to a highly developed city in a large industrial economy, but the country can readily rebuild it with its own resources.

Tornadoes and hurricanes can be regarded as discrete events in distinction from broad and long-term climate shifts. But another intermediate range phenomenon exists as well. Regional climate can shift for a few years at a time due to changes in the jet stream. During the great depression of the 1930s, a southward shift in the jet stream created a regional drought that devastated the agricultural output of the central prairies. Not only did this severe climate event influence the nation's economic output and the course of its recovery, but it also displaced millions of residents and destroyed their capital base (both financial and physical capital). Being part of a much larger country meant that these residents had the opportunity to relocate and rebuild their lives. Not only could the afflicted region draw upon financial resources from the federal level, but the country's geographic scale provides a physical

DOI: 10.1057/9781137340313

escape without international border constraints. Size matters. If instead the region were an independent nation, the economic impact would have been overwhelming.

The most important feature of larger countries is that they are much more likely to encompass multiple climatic regions. So a unique type of opportunity for economic development will exist within larger countries: the movement of economic factors within the country. In a large area under one set of rules, the movement of people, goods, and capital can carry on quite efficiently. The effect is that large countries can optimize their economies within their own borders. There is great opportunity for internal trade and specialization as well as the resulting economies of scale. And the result is almost universally a concentration of capital and populations in the most temperate climate areas. This is very easy to see in economies around the world such as China, Brazil, the US, Canada, Greenland, Argentina, and Australia.

The 'institutional' view holds that a country is a political entity formed of laws, customs, social mores, and the like. Advocates of this view of development cite the economic differences between geographically identical pairs of countries such as Haiti and Dominican Republic, or North Korea and South Korea. They argue that if climate was the most relevant factor, these states should be much more similar because they are physically adjacent to each other. One aspect of that argument holds true: destructive institutions like corrupt regimes can result in disparate economic performance within a climatic region. But these countries have small-scale and short-term differences in wealth, nothing like the permanent chasm between the former East Germany and Laos. Furthermore, the institutional view has nothing to say about why the North–South development pattern in Africa mirrors the pattern within Chile or within Brazil. Nor can it explain the symmetry of global development about the equator. Finally, it seems absurd to suggest that legal institutions are arrayed in the same manner from the equator toward *both poles*, without admitting that geography plays a fundamental role.

Only climate can address economic issues on the largest scale. As to the North–South issue, nothing in the climate explanation of long-term development suggests that poorer countries cannot improve. They can improve their economic results, and the Southern Asian regions have shown one path to that end. Closing the gap with the economically advanced and dynamic northern countries remains another question altogether.

DOI: 10.1057/9781137340313

Notes

1 See http://www.nsi-ins.ca/ (Accessed January 15, 2013).
2 See http://unstats.un.org/unsd/snaama/dnlList.asp (Accessed January 15, 2013).
3 Antigua and Barbuda, 1981. Bahamas, 1973. Barbados, 1966. Dominica, 1978. Grenada, 1974. Jamaica, 1962. Saint Kitts and Nevis, 1983. Saint Lucia, 1979. Saint Vincent and the Grenadines, 1979. Trinidad and Tobago, 1962.
4 J. De Gregorio; *Economic Growth in Latin America: From the Disappointment of the Twentieth Century to the Challenges of the Twenty-First*, Central Bank of Chile Working Papers, No 377 (abstract), November 2006.
5 http://data.worldbank.org/indicator/SI.POV.GINI

DOI: 10.1057/9781137340313

7
The Isolated Islands

Abstract: *Small island nations offer the sort of natural experiment that economists prize. The Caribbean and Oceania are presumed to be subject to high prices and heavily reliant on trade. The structure of small island trading economies might be expected to increasingly resemble their closest and largest neighbors. Instead, they closely resemble countries with the deepest cultural connection. Their industrial structure is related to the Gravity Theory of trade.*

A comparison of Cuba and Australia shows how totally different political systems do not distort the relation of agriculture to wealth.

Key words: Gravity theory of trade; isolated islands; trade theory

Atikian, Joe. *Industrial Shift: The Structure of the New World Economy*, New York: Palgrave Macmillan, 2013. DOI: 10.1057/9781137340313.

The economist's prize

Small island nations present the sort of natural experiment that economists prize. The Caribbean and Oceania are presumed to be deeply reliant on trade because of their isolation, and to suffer high prices because of small scale and the long distances to their export markets. As a result, they should also undergo slower industrial progress due to the added difficulties involved in communications and international technology sharing. This idea can be tested by comparing their structure with the continental regions, and by using the concepts found in the Gravity Theory of trade. The surprising results are that the island groups fare no better or worse than the continents, while size and distance matter little to structural progress.

Figure 7.1 charts the familiar pattern of falling agriculture and rising wealth in the world's major regions, including Oceania and the Caribbean.

To help visualize the geographic trend, a curve was fitted to the charted points representing the regions. The resulting curve starts with the familiar combination of high agricultural share and low GDP per capita in the lower right hand corner. The Asian and African tropics predominate here. The curve extends toward the high output economies of North

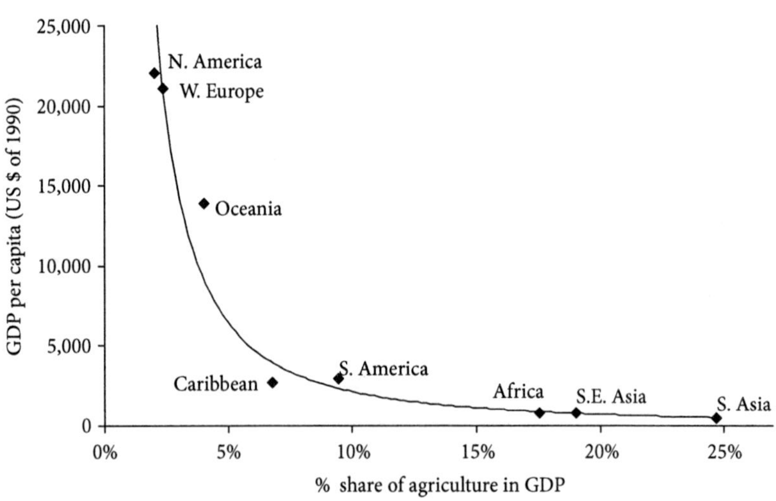

FIGURE 7.1 *World regions, agricultural share vs. GDP per capita, 1970–2008*

Source: Derived from UN National Accounts Database.

DOI: 10.1057/9781137340313

America and Western Europe in the upper left hand corner. These are the temperate climate nations heavily biased toward the services sector.

Oceania and the Caribbean present an unexpected result. They occupy the central area of the chart, suggesting that they have no significant bias in their structure and wealth compared to the continents. In other words, some continents are wealthier than the islands while some are poorer. The islands don't suffer an economic disadvantage due to small scale, or a technological setback due to isolation. Why does this result conflict so directly with the seemingly justified expectation?

Being a group of island nations usually implies a reliance on trade and, in turn, the economic disadvantage of high transport costs. Trade economists would expect these small or isolated economies to trade primarily with the nearest large economy. Their industrial structure may also be prone to eventually resemble those large nearby economies. But Oceania and the Caribbean clearly do not. Instead, each island group's agricultural share and income level differ greatly from the closest and largest neighbors (China and the US). Oceania's point on the chart lies closest to Western Europe while the Caribbean's point lies closest to South America.

Oceania is a far flung group of tiny islands with small economies, but is vastly dominated in size by Australia. This means that the group's position on the chart is highly biased by one industrially advanced country. Meanwhile, the Caribbean economies are more uniform in size, but they lie directly between North and South America so that the effect of proximity is balanced. A closer look at two country examples is in order.

Figure 7.2 charts the industrial progress in Cuba and Australia over recent decades. Among all of the island nations, these were chosen on account of their vast differences. Size, location, language, history, and, most especially, political systems, could hardly be more distinct. So, if agriculture and income in these disparate countries follow the regional and global patterns, it would support the idea that industrial structure is a universal indicator of development. This is exactly the outcome.

Australia and Cuba each showed rising GDP per capita over the years, but Cuba surprisingly produced the faster increase (144% vs. 95%). Owing to the scale of the graph, it is difficult to see that Cuba's GDP per capita doubled by 1984. Both countries focused more intensively on mining, manufacturing, and services. Their farm sector share fell at the typical rates, rapidly for poorer Cuba and slowly for wealthier Australia. Of course the level of wealth is vastly diverse between them, but that is

DOI: 10.1057/9781137340313

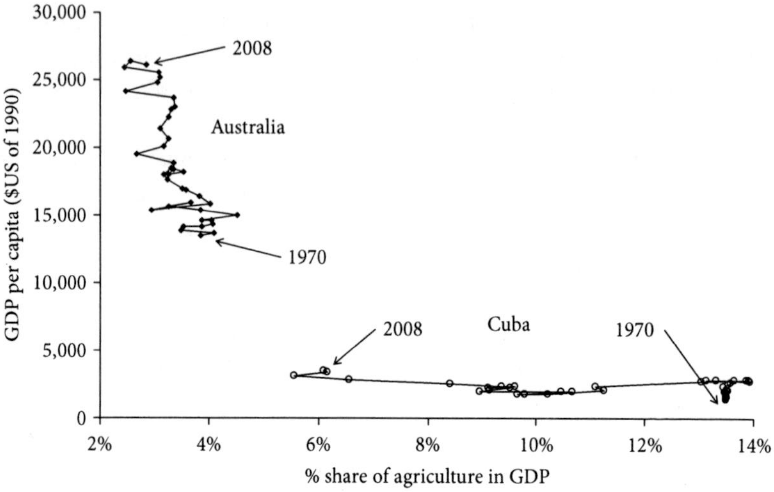

FIGURE 7.2 *Australia and Cuba, agricultural share vs. GDP per capita, 1970–2008*
Source: Derived from UN National Accounts Database.

a common feature when comparing dictatorships with open economies. Perhaps the most remarkable aspect of this comparison is that regardless of the fifty-year US embargo on trade with Cuba, its industrial transition has gone ahead unabated (see the Box on Political Structure).

Focus on political structure

Comparing open economies with dictatorships invites the question of whether or not political systems play a key role in determining growing prosperity. Some institutional thinkers believe that central bank independence is a key indicator of openness and of wealthy economies, but most of these states were wealthy well before central banks existed. What about democracies? Democratic nations are widely believed to rank high in GDP per capita. But is this true?

Many studies of wealth do not find that the government type correlates with wealth. For example, notice that India is a democracy that is not wealthy, and in fact is quite poor. Well, it has not been a democracy for very long, about 60 years, which may not be long enough for wealth to develop. So does wealth depend on long-term democracy? That won't work either because most countries at the top of the wealth list are not democracies, but tiny monarchies

DOI: 10.1057/9781137340313

or territories with appointed leaders. The larger Euro states were monarchies for centuries before changing to electoral governance in modern times. Sweden is not a long-term democracy, but rather a monarchy until 1917. The same applies to top ranked Norway and Luxembourg. Highly ranked Finland was a Russian territory until the First World War. Japan is as wealthy as the US, but it became democratic only in 1947. And what about the very top of the list? Monaco and Liechtenstein have never been democracies, but monarchies.

Claiming that wealth depends on democracy won't work, and neither will long-term democracy. Even so, dictatorships do not figure prominently in the high wealth category, nor do democracies at the low end. So the feeling remains that political systems do have some bearing on economic standing. That may be true to some extent, but there is one extreme bias that sows confusion. The bias is that dictatorships never produce wealthy countries, while democracies sometimes produce poor ones and sometimes produce wealthy ones. As a result, predicting growth based on political systems simply doesn't succeed across the entire range.

A powerful attraction

The Gravity Theory of trade provides another perspective on the isolated island groups.[1] It treats countries as if they were planets in mutual attraction. Simply put, larger closer planets exert more gravitational pull than smaller distant planets. Likewise, a large nearby economy should attract more trade than a small distant economy. In other words, the amount of trade between two countries depends on size and distance. For instance, tiny Viet Nam would be expected to trade most heavily with the enormous Chinese economy which lies right next door, as opposed to tiny, distant Peru. The gravity theory is regarded as very successful in predicting trade quantities, but so far it has not addressed industrial structure. This raises the additional question of whether economies advance in tandem based on their size and distance.

Gravity affects trade, so, does trade in turn affect industrial structure? The isolated islands provide a perfect experiment. Does the structure of an island trading economy develop to increasingly resemble its closest and largest neighbors? For example, does Australia's structure resemble that of Indonesia, China, and India? Does Cuba's structure resemble that

DOI: 10.1057/9781137340313

of the US? It turns out that economies do not resemble each other as might be expected based on size and distance. Instead, their structure more closely resembles those with greater cultural connections.

The largest Caribbean island economies are primarily Spanish speaking, and as a group, their structure closely resembles South America. This situation may not seem completely consistent, because of a significant exception, Jamaica. The economic structure of English-speaking Jamaica also matches the South American structure, so resemblance based on cultural background appears to break down. Nevertheless modern gravity theory would point to Jamaica's Spanish origins and conclude that its structural resemblance to South America is unsurprising. Cultural connections are thought to be extremely persistent, their effects lingering for centuries.[2]

Does Cuba's structure resemble South America's more closely than the US because it was held back by the embargo? This seems unlikely because it also closely resembles many of the other non-embargoed islands, as well as many of the non-embargoed Central American economies. Instead, Cuba's economy is perfectly typical of those in its region and its shared cultural background. Its resemblance to neighboring states does not prove that the embargo had no effect on trade, though. One such effect stands in a complete opposition to gravity theory; Cuba's main trading partner in recent years has been the Netherlands instead of the more likely Brazil or Venezuela.[3] This strange case in which a small, distant, unrelated country ranks first in trade may have arisen as an unexpected response to the embargo, perhaps by sprouting a transshipment industry.

Australia conducts one-half of its bilateral trade with OECD countries. Such a large percentage at first seems strange given that the OECD group is geographically the most distant.[4] Although this fact points to the later discovery that cultural factors also determine trade volume, it may suggest a more significant link, namely the effect of culture on industrial structure.

One indication of the gravity theory's early crudeness involved the way it addressed large countries or bordering countries. For instance, Nassau Bahamas lies close to Miami, USA, and much of its economic connection stems from this short distance (192 miles). But the theory assigned the distance between the countries as the distance between their capital cities. As a result, Nassau's distance from the US is deemed to be the far longer distance from the island to Washington DC (1100 miles). This

DOI: 10.1057/9781137340313

contributed to inaccuracy in prediction as well as some hesitation in adopting the theory. The case of a shared border such as Canada and the US created the absurd idea that the distance between the countries was zero, while using the capital cities did nothing to account for the reality of various regional markets along the border. Later refinements solved this distance problem by developing the idea of 'economic distance' that accounts for all barriers instead of only transport cost.

A large economy is expected to exert a tremendous influence on a nearby small economy; the small one will merely exert a weak influence in return. This implies that a small country could gain some share in its large neighbor's market by exporting similar goods. As a result, this trade flow could help drive the structure of the smaller economy to more closely resemble the larger one. Note that this result is not currently considered to be a part of the gravity theory; it is proposed here for the first time. (See the box on Gravity theory below.)

Some economists conversely believe that the smaller country would develop a different industrial structure due to specialization, although the experience in North American or European regions does not uniformly support this idea. In this view, Canada, for example, would specialize its exports, perhaps in agricultural products, and so end up with a very distinct industrial structure relative to the US. Instead, Canada's small economy, deeply reliant on trade with the US, has a nearly indistinguishable economic structure. Agriculture contributes a small share of GDP, and output per capita is high in both countries. Canada has no large nearby trading alternative, so another example such as the Caribbean offers an opportunity to confirm the idea.

Geographic distance suggests that the Caribbean receives an equal influence from North and South America (owing to its position about midway between the continents). The small-scale island economies should benefit from opportunities to specialize and transport goods just as easily in either direction, north or south. However, the vastly larger US economy should be the dominant influence and bias the islands' GDP per capita higher than South America's. This suggests that in Figure 7.1, the Caribbean should lie closer to North America than to South America.

The expected result has clearly not occurred, as Figure 7.1 shows the Caribbean output level and structure to closely resemble South America. The simple and obvious implication is that cultural association plays a much stronger role than geo-economic factors of GDP scale and physical distance. This can also be confirmed by examining Oceania's situation.

DOI: 10.1057/9781137340313

Oceania is economically dominated by Australia and, secondarily, by New Zealand. These islands are geographically isolated and should bear a disadvantage due to relatively high shipping costs for their traded goods. Trade should tend to favor nearby economic giants in Asia. As a result, Australia's industrial structure should not be expected to evolve to resemble Europe's. Southeast Asia, being the adjacent economic group and the world's fourth most populous region, might also have biased Australia's structural away from Europe's, but neither of these expectations materialized. Instead, Oceania lies at the opposite end of the graph from Southeast Asia (Figure 7.1). Again, it seems that the obvious cultural affinity of Australia and Western Europe has outweighed the geo-economic factors.

Gravity theory started as an explanation for trading patterns, and has since been extended to other fields including business and finance. Now, industrial structure creates an additional concept within the theory. That is, the gravitational attraction between countries with a shared culture tends to generate similarity in their industrial structure. So, for example, Oceania's industrial structure stems from Australia's historic association with Western Europe, more than any geographic factor. Likewise, the Caribbean's structure gravitates toward South America's structure due to its deep historic relations.

As long as culture is considered to be a part of gravity theory, it can be used to help understand the development of industrial structure. Structure, however, does not seem to stem from trade. These findings do not work to refute the gravity theory of trade, as it was not originally intended to address industrial structure. But another level of geo-economic analysis is now possible.

Focus on gravity theory

The gravity theory of trade says that the influence exerted by trading economies upon each other resembles the force of gravity in physics. The force of gravity exerted by an object is proportional to its mass and inversely proportional to the square of its distance to another object. Likewise in trade theory, the amount of trade is analogous to gravitational force, while mass is analogous to GDP. In the theory's earliest form, 'distance' simply referred to the geographic distance between the countries as an easy way to account for the effects transportation cost.

DOI: 10.1057/9781137340313

The earliest versions of the theory produced only moderate correlations between distance and trade volume. One of the underlying problems was that the distances used were between national capitals regardless of how relevant these cities were to the movement of goods. One solution leading to an improved correlation was the concept of 'economic distance', as opposed to the natural geographic distance. This concept incorporates all of the barriers to trade within the distance term. Key among these various barriers is a term called "culture." Culture captures the effects of long-term traditions, otherwise referred to as habits, customs, colonial relationships etc. Differing culture between two countries diminishes trade whereas shared culture enhances trade.

The correlation between distance and trade improved, but the entire concept of gravity has suffered a conceptual break. This revised theory of trade that includes culture no longer resembles gravity, because gravity is uniquely about size and distance. Adding a cultural term to economic gravity is analogous to adding a language term to an equation in physics. Planets would then be said to orbit their sun as a function of mass, distance, and whether or not the predominant language was English. The theory including its cultural aspects can still be functional and true, but it should fall into a different kind of category than gravity.

Notes

1 The word "theory" is most often used to imply guesswork or an abstract idea unrelated to practical reality. Instead it should be used to mean an explanation for the way that things behave. In this case, the gravity theory is simply an explanation for the amount of trade between countries based on some of their physical features.

2 For a detailed treatment of cultural persistence, see D. Campbell; *History, Culture, and Trade: A Dynamic Gravity Approach*, UC Davis Working Paper, 2010.

In Puerto Rico's case, its full economic integration with the US renders it irrelevant to this issue because it functions as a US state instead of an independent trading nation.

3 Cuba's four main trade partners in 2006 were Netherlands, Canada, Venezuela and China. See Cuba's statistical service; Oficina Nacional de Estadisticas www.one.cu/aec2006/anuariopdf2006/capitulo7/VII.5.pdf

4 See Australia's Department of Foreign Affairs and Trade http://www.dfat.gov.au/publications/tgs/index.html

DOI: 10.1057/9781137340313

8
Farms and Factories

Abstract: *Productivity improvement through technological change is the key cause of industrial shifts. Farms have continually improved labor efficiency, leaving surplus labor to shift to manufacturing. Innovative factory products cycle back to the farm in an ongoing cycle. Cause and effect are permanently intertwined.*

Agriculture has nearly vanished as a share of GDP, but its output always grows. Manufacturing has a moderate and stable share of GDP, but is unevenly distributed around the world.

Shifts from farm to factory to office are blamed for job losses. Trade with China is also blamed. These are only partly correct. China regained its most favored nation trade status and is now in an industrial revolution. It has added to global industrial output instead of absorbing it from the West.

Key words: China industrial revolution; innovation; productivity improvement; technological change

Atikian, Joe. *Industrial Shift: The Structure of the New World Economy*, New York: Palgrave Macmillan, 2013. DOI: 10.1057/9781137340313.

 DOI: 10.1057/9781137340313

Industrial shift: the cause

How does industrial structure shift? Far from being a stagnant sector, progress in agriculture equals or exceeds the other sectors. But more significantly, this progress does far more than improve farm output. Progress on the farm launches the entire cycle of industrial shift.

The shift away from agriculture has its roots in productivity improvements that reduce the number of people needed in the sector. From the earliest stages of industry, technology has raised productivity following a general pattern that makes each aspect of farming easier and less costly. Human labor is augmented by adding physical power, first from animals, and then by mechanization. Land is augmented by irrigation, and then much later by chemical fertilizers. Plants and animals are augmented by selective breeding, medicines, and genetic manipulation. Finally, knowledge is augmented by record keeping, mathematics, and by new technologies such as GPS and satellite remote sensing. Each person produces more; each plot of land produces more; each dollar invested produces more. Overall fewer people and resources are needed to feed the population, so more resources shift to meeting other needs and wants: medicine, education, housing, transportation, security, entertainment. (Note that shifts in agriculture alone do not cause output growth, because there are also simultaneous shifts in the other sectors. For example, shifts in agriculture already incorporate the productivity improvements found in manufacturing and services. When factories adopt labor-saving equipment, information systems, and new scientific methods, farms adopt similar technologies. So, changes in each sector cause changes in the other sectors.)

What kind of the evidence supports the idea that productivity drives the industrial shift? Employment provides the main clue. The record of US farm employment, for example, shows that consistently fewer jobs exist in the sector every decade since 1900. At the same time as farm employment decreases, farm output consistently increases. This is the very definition of increasing productivity.

While agriculture plays a continually diminishing role as a share of global GDP, both the level and growth of farm productivity are robust and clearly evident in national records. For example, from 1970 to 2003, Sweden's farm labor productivity outstripped manufacturing by up to three times. Likewise, a broader sampling of OECD countries shows that agriculture leads manufacturing in total factor productivity growth in most countries (see Appendix B for the country list).[1]

DOI: 10.1057/9781137340313

Unlike the farm sector, the change in the manufacturing share of global GDP is nearly imperceptible. Not only did its share move by less than 1.5% over the past 40 years, but most of this change was due to a single isolated event, the 1990 collapse of the USSR. The loss amounted to nearly $300 billion in manufacturing output per year, without which the sector's share in world GDP would have registered a slight increase. The successor states never recovered their Soviet-era manufacturing output, meaning that the rest of the world's manufacturing share has remained stable.

Many recently advancing economies have a growing share of GDP from manufacturing (Botswana, South Korea, India, Mexico, Belarus, Bolivia) while the most mature industrial states lead the worldwide shift away from manufacturing and toward services. For example, the US manufacturing sector share of GDP has gradually but steadily shrunk from 28% in 1970 to 22% in 2008 and it appears set to decline further through the current recession and beyond. In effect, manufacturing share has passed its peak in the wealthier countries while it approaches its peak in others.

As with farming, manufacturing undergoes continuous change. Its share of global GDP has held steady for decades but other aspects are in flux: output continues to grow; improved productivity has affected employment; innovation changed the kinds of technology being produced; and commercial pressure has redistributed operations around the world. Just as farms were the first step in the industrial shift, factories are the second step. Factories absorbed former farm workers, but factories have also increased their productivity, and in turn have lost workers to the ever growing services sector.

Now industrial progress or advancement can be more clearly defined at a general level for national economies. Progress can be understood as the ongoing shift in concentration from older sectors to newer sectors and new goods, without shrinking either sector. So agriculture grows its level of production while manufacturing grows faster. And manufacturing grows while services grow faster still. While this forms the general trend around the world, special cases do arise. Agriculture can raise its share of the economy and still constitute progress. For example, Brazil and Chile have recently done this by adopting technology to increase their farm productivity. Its people are better fed while the county's food exports increase. Meanwhile some of Brazil's rising farm output goes to producing bio-fuels to decrease the country's reliance on crude oil as an energy source.

DOI: 10.1057/9781137340313

The Western world has recently undergone a critical change in the relationship between farming and national income. Prior to 1950, the farming share of the economy fell quickly as the great shift toward factories geared up. The bigger the factory sector grew, the more income the nations generated. By the 1970s, farming was down to about 3% of GDP and shrinking its share only gradually. GDP per capita did the exact opposite, rapidly accelerating.[2] In other words the structural shift from agriculture had virtually stopped while national wealth continued to shoot higher. This may be the clearest indicator of an economy being in the ultimate stages of industrial development, and it occurs in virtually every advanced Western country. It remains a fact that agriculture is a very highly mechanized and productive activity in these countries, but the past few decades mark the first ever period in which agriculture forms a vanishingly small share of the wealth-producing capacity of industrial economies. This, the oldest industrial sector, can be expected to be the smallest because technological change has been raising its productivity for the longest time.

Industrial shift: the effects

While nobody wishes for a return to the technologies of a century ago, change does bring fear of displacement and deprivation. Millions of people appreciate not having to shovel coal or harvest wheat by hand, but they also fear losing their factory job to further automation, even though their own jobs were the result of automation on farms and in mines. People are also understandably confused; on a nationwide scale, millions of farm jobs have been lost over the decades while millions of factory jobs have been added; within the manufacturing sector, textile factories have been lost while auto factories have been added; within the foreign trade sector, goods are in deficit while services are in surplus. How does it add up?

The popular view that trade with China reduces US employment seems to be unfounded in the facts. It also turns out that the impression of a general decline in farming and manufacturing is unfounded. There is no doubt that these sectors have changed a great deal, but they have not declined.

According to an increasingly common view, jobs, exports, and manufacturing output all decline together. Government and corporations take

DOI: 10.1057/9781137340313

the blame for creating the conditions that cause these shifts, even though consumers have proven to prefer low priced Chinese goods while arguing in favor of high-paying local manufacturing jobs. The financial crisis of 2008 reinforced the blaming because the huge number of recent job losses has combined with two already highly publicized trends: the government's rising trade deficit, and the corporations' off-shoring of factory jobs. The claim is that profit seeking firms move their factories to lower wage countries, resulting in lower domestic employment and production. Then rising imports from relocated factories replace domestic goods and cause a larger trade deficit. This mechanism has not been the case in the US as outlined in the following three points.

First, manufacturing output has not declined. Instead, it has increased by more than 100% since 1970. As the services sector accelerated, there has been a decline in the *share* of manufacturing in GDP from 29% to 22%, but this started decades before trade with China, and decades before the US even had a trade deficit (see Figure 8.1). Again, the manufacturing share has fallen but the total amount manufactured has risen. Little question exists as to why the common perception is of manufacturing collapse. Most makers of clothing and household goods have indeed moved overseas, and these goods have highly visible labels showing the country of origin. Ten million consumers can easily see that a shirt or toaster was

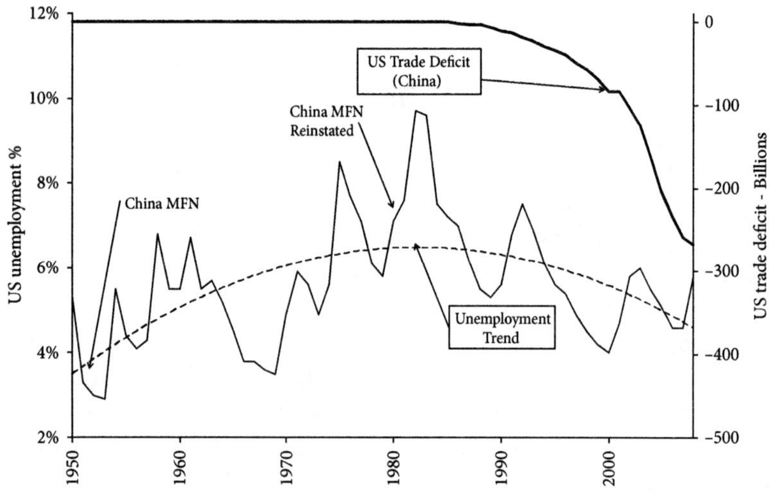

FIGURE 8.1 *US–China trade deficit vs. job trend, 1950–2008*

Source: Derived from US Bureau of Labor Statistics and US Department of Commerce.

DOI: 10.1057/9781137340313

made in China or Indonesia, and conclude that imports cause the daily bad news. The origin of other kinds of goods cannot be so easily seen by the millions of people that use them. What goods are manufactured in the US? Here are a few examples: Boeing aircraft; Harley Davidson motorcycles; John Deere tractors; Caterpillar bulldozers; the NASA Space Shuttle; GE MRI machines; Wabtec rail equipment; Ford automobiles; Toyota automobiles; BMW automobiles; Sea Ray boats; Sub-Zero refrigerators; Cummins diesel engines; Kenworth trucks; Hansen bricks; Illumina genetic sequencers.

Second, the US jobs trend has moved in the opposite direction from the decline expected due to the trade deficit. In 1951 the US suspended China's Most Favored Nation trade status. In Figure 8.1 the dotted line shows the trend in unemployment worsened during the no-trade period that lasted until 1980.

Job losses must have been caused by something other than trade. China's MFN status was restored in 1980, and the surprise is that the US job trend turned positive just as trade with China increased. The trend improved by about 2% since large-scale trade with China and overall trade deficits resumed in the early 1980s. The unemployment trend fell from about 6.2% to 4.3% from 1980 to 2010, so the US trade deficit in manufactured goods ran counter to employment. A higher trade deficit was associated with *lower* unemployment.[3]

Although a full discussion of the causes of unemployment goes beyond the present scope, one seldom heard factor that connects jobs and trade should be mentioned. The world price of crude oil has a strong connection to US post-war unemployment.[4] The off-shoring of mass consumer goods came partly in response to higher oil prices starting with the 1973 and 1979 OPEC crises. Why? Expensive oil makes everything more expensive, from food to clothing, and home heating to driving. Rising oil prices brought inflation and job losses, but consumers struggled to change their high cost activities by switching to new cars or home heating fuels. These require a large cash outlay or increased debt. But it was very easy to buy lower priced clothing and shoes, which retailers could quickly source from overseas. Permanently higher oil prices helped to push manufacturers of textiles, clothing, shoes and furniture to lower cost locations. To the extent that some firms moved their factories, more jobs were also displaced. Like off-shoring, domestic labor productivity also contributed to lower manufacturing employment. This is discussed in the next chapter.

DOI: 10.1057/9781137340313

Third, some economists studying this question agree that the trade deficit is not likely a result of off-shore sourcing or of China's exchange rate policies. For example, in 2000, the United States Congressional Budget Office concluded that a long decline in domestic personal savings was the main cause of the trade deficit. The CBO report also concluded that trade with China and all other countries was an irrelevant factor.[5] This kind of conclusion often draws harsh criticism and accusations of political bias, with claims that such reports are covertly intended to justify the government's policies. A brief look at other countries, however, can help to confirm the report's findings. Japan, for example, has a very high savings rate, and a trade surplus. China also has a high savings rate and a trade surplus. More interestingly yet, these two countries have vastly differing currency exchange rates: Japan's is high while China's is often thought to be too low. Similar signals come from Canada and Germany, where high savings and trade surpluses again suggest that the exchange rate is less relevant than often thought.

~ ~ ~ ~

While the general trends show rising output in all three industrial sectors globally, progress causes pain as people shift between farms, factories, and offices. Remedies are often proposed in order to reduce unemployment and trade deficits that are perceived to be related to these shifts, so a few challenging questions should be asked. Will higher exports and a trade surplus reduce unemployment? Will lower taxes lead to more manufacturing? If goods are bought overseas, do local industries get stranded in other less advanced activities such as logging or fishing? Some unanticipated facts lurk in the data.

Unlike the US, Germany turned out a hugely positive trade balance that would stir the Western world's envy. This trade surplus amounted to about 7% of GDP, which raised expectations of much lower unemployment. Instead, unemployment reached its all time peak of 11.1% even as the trade surplus soared by the mid 2000s. So export strength was no cure-all for Germany, just as export weakness was no disaster for America. In reply it could be argued that Germany's high unemployment was caused by the country's 1990 re-unification which brought low-skilled East Germany into the economy. This rationale does not hold up though, because the rising unemployment trend began twenty years prior to re-unification and trade held steady in that period. In general,

DOI: 10.1057/9781137340313

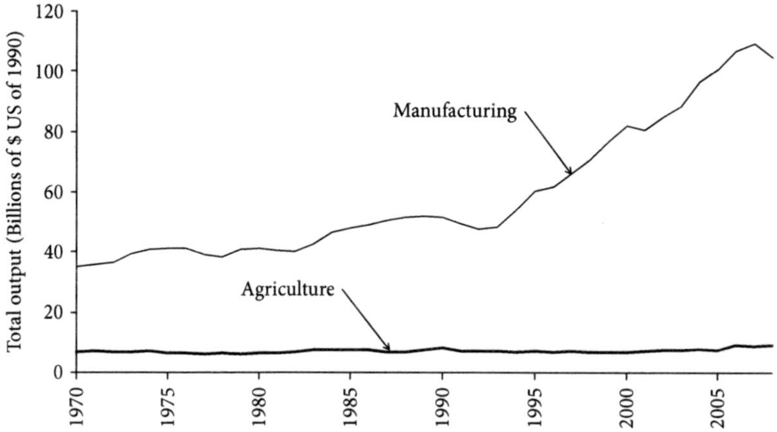

FIGURE 8.2 *Sweden two sector output trend, 1970–2008*
Source: Derived from UN National Accounts Database.

trade balances and manufacturing shares remain unrelated to economic strength and employment levels.

Sweden and the US, two wealthy democratic industrial economies, also provide an illuminating comparison due to their significant differences. Sweden's economic scale pales in comparison, at only 1/20th the size of the US in real GDP. Sweden is also famous for its high taxes and government intervention in the business sphere. So how does its industrial structure compare to America's? It has a very similar outcome. Figure 8.2 shows its low and stable agricultural sector, and growing real manufacturing output. Both countries display a structure typical of highly advanced economies worldwide.

Manufacturing marks a key difference between the countries, where highly taxed Sweden has outperformed. Sweden's recent manufacturing output growth of 190% since 1970 fairly exceeds America's 140%. Furthermore, Sweden's manufacturing share in GDP has grown by 7% whereas America's has fallen by 7% over recent decades. This puts Sweden in the minority of highly developed economies with a rising trend in manufacturing share, but it also casts serious doubt on claims by some US observers that lower taxes will necessarily create new manufacturing firms and factory jobs.

Canadians famously criticize themselves for being overly reliant on resource extraction and raw materials that hold little value. These are seen as low skill and low value sectors that earn its citizens the belittling

DOI: 10.1057/9781137340313

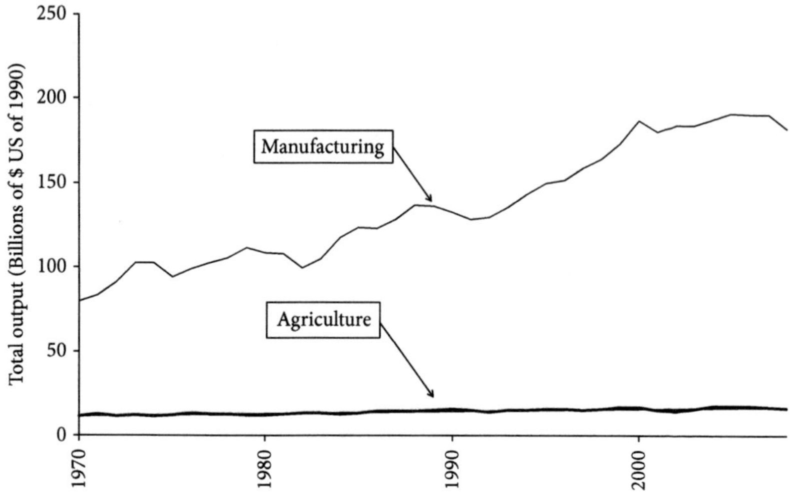

FIGURE 8.3 *Canada two-sector output trend, 1970–2008*
Source: Derived from UN National Accounts Database.

description, 'hewers of wood and drawers of water'. More importantly, the exporting of raw materials is thought to suppress manufacturing and its related jobs. Perhaps this view persists because of the intense attention focused on failing firms and the low profile of growing firms, but the facts do not justify the self-criticism.

Canada's farms and factories are clearly not part of a receding industrial sector. Figure 8.3 shows that annual agricultural output has grown by $4.7 billion since 1970. In the much larger manufacturing sector, annual output has grown by $102 billion.

Furthermore, Canada's manufacturing output grows at more than double the rate of its population.[6] This reflects rising labor productivity as well as the gradual shift in industrial structure toward the services sector. US figures are nearly identical for the period.

As a share of GDP the manufacturing sector declined from 27% to 20%, which is indistinguishable from the majority of countries worldwide, and in line with many advanced economies, within the OECD such as Denmark, US, Norway, and France. Figure 8.4 shows the OECD country rank for manufacturing share.

The perception that Canada's economy relies on exporting raw materials comes in part from the high level of mining expertise in the country. More mining companies are headquartered in Canada than in any other

DOI: 10.1057/9781137340313

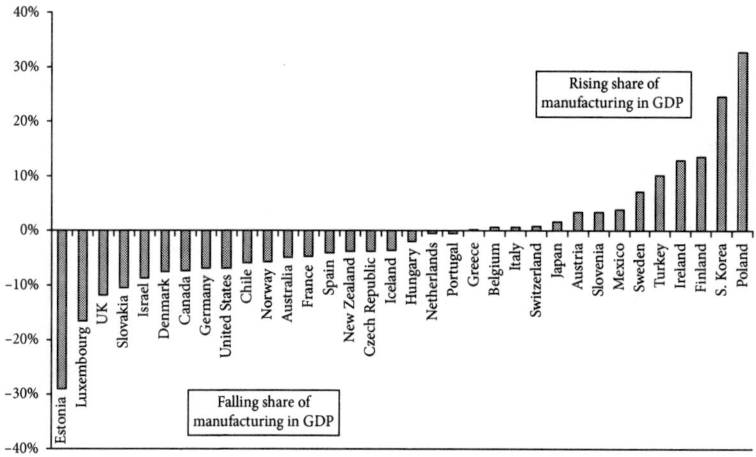

FIGURE 8.4 *Total percentage change in manufacturing share: OECD countries, 1970–2008*

Source: Derived from OECD Annual Labour Force Statistics Database.

country, including 12 of the world's top 100, which is quite remarkable for an economy with only 2.5% of global GDP. But in case that gives the impression of mining as a huge employer, it shouldn't. If it seems that Canada has become a so-called petro-state, it hasn't. Petro-state status belongs to Saudi Arabia or Trinidad and Tobago, whose oil sector makes up nearly 50% of the economy as opposed to Canada's 5%.

Although oil and gas production have increased over the decades, and their prices affect the Canadian dollar exchange rates, the sector does not top the list in output or in jobs by a wide margin. The number of jobs in mining, oil, and gas extraction combined remains at 1.7% of total national employment, and has been virtually unchanged since at least 1976.[7] Manufacturing employment is six times larger than mining. Finally, for 2012 Statistics Canada reports that forestry and logging make up only one-third of one percent of Canadian economic output, which entirely refutes the idea of a Canadian dependence on raw materials.

So if manufacturing is not in decline, what does Canada produce? Several product categories have indeed been nearly eliminated in recent decades, including textiles, shoes, furniture, and clothing. But as Figure 8.5 shows, these have been replaced by autos, aircraft and satellites, trains, metal products, food, chemicals, and machinery. These are the top five categories that make up 60% of manufacturing output.

DOI: 10.1057/9781137340313

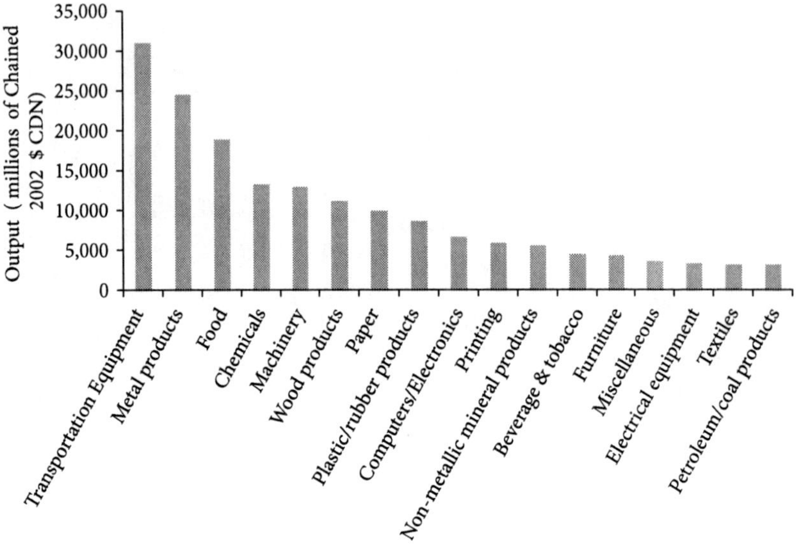

FIGURE 8.5 *Canada manufacturing sub-sector output ranked by value, 2008*
Source: Derived from Statistics Canada GDP Annual Table 1–3.

The trend is toward producing more complex and technically advanced goods, with more added value instead of the popular perception of less added value. In contrast, wood products and paper are the sixth and seventh ranked manufacturing categories, each comprising about six percent of the sector total or a mere 1.2% of GDP. "Trains, planes, and automobiles" gives a far better picture of Canada's economy than "hewers of wood". "Financiers" would be better yet.

The Eastern industrial revolution

China's current industrial transition compares in some ways to a previous global shift. In 1900 Britain was poised to lose its preeminent position as the world's largest economy to a country with which it had been deeply tied for centuries. That country was the United States, which, like today's China, was also an agricultural economy with a burgeoning manufacturing sector. The US would take its global lead with enormous waves of immigration and new workers from the old world. Today's situation has parallels and contrasts. Today's industrial and economic shift obviously centers on China, still agrarian to a relatively large extent, but also with

DOI: 10.1057/9781137340313

surging manufacturing. The difference is that China is absorbing three simultaneous industrial waves in the form of foreign capital, foreign technology, and an internal workforce migrating from countryside to city, literally a transition from farm to factory.

Some effects of a major global transition are well known from the American episode. A shrinking labor pool in Europe raised wages and living standards; new manufactured products improved America's farm efficiency, leading to more urbanization; rising financial clout in New York diverted power away from Britain. Comparable effects may come from China's internal shift, the equivalent of bringing another large country onto the world economic stage.

When America first welcomed a sea of strangers in the 19th century, it did not change the size of the global workforce. The new Americans were simply relocated, mostly from Europe. In China's case, 300 million people are in effect being introduced as new workers on the world stage. They already existed, of course, but operated as an internal farm economy, closed to the world. There is literally no precedent for the sudden infusion of factory labor into the world market, and the effects have yet to be fully seen or understood. Economists and journalists recognize that Western middle class wages have stagnated over recent decades, but admit that the cause remains unclear. Could it be that by adding a new workforce nearly equal in size to the entire US population, China has already started to change the global distribution of personal income? The final chapter takes up this question.

China still shows some clear contrasts to the industrial structure of the highly advanced economies. It is not the case that China has little or no advanced industrial capability, but that a much greater share of its economy still lies in agriculture. Fully 10% of China's GDP derives from agriculture whereas Sweden's has recently dropped below 3%, and America's has long since stabilized at 2%. This does not, however, imply that China lags the West in the same way that it did before the 1970s. China's distinguishing feature today is not only its farm-heavy structure, but the speed of its ongoing shift from the farm to the factory.

While many highly advanced economies have shown a gradually declining manufacturing share, falling less than 10% since 1970, China's share has *increased* at double that rate. Likewise, instead of America's stable 2% agricultural share in GDP, China's share has plummeted from 52% to 10% even as its real farm output quintupled. The only reason that its agricultural share has dropped so much is that manufacturing

DOI: 10.1057/9781137340313

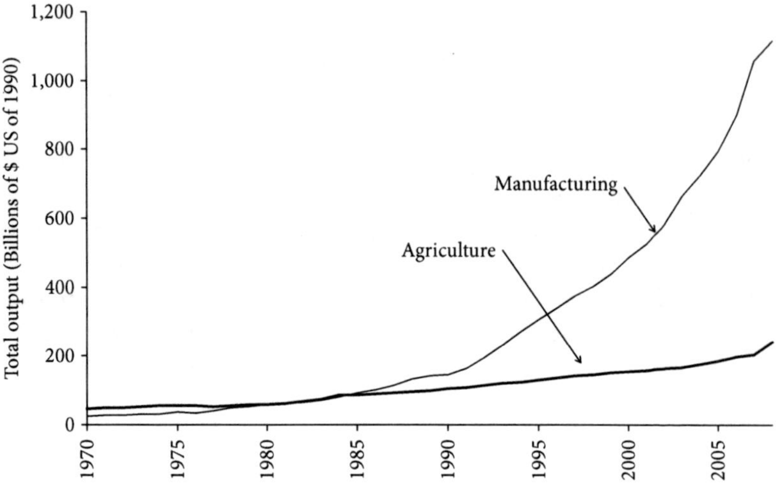

FIGURE 8.6 *China two-sector output trend, 1970–2008*
Source: Derived from UN National Accounts Database.

output has grown nine times faster. Figure 8.6 shows that factory output has grown 46-fold since 1970, unmistakably signaling China's industrial revolution.

The most recent annual data, for 2011, marks another large increase in China's factory output. Depending on how inflation is accounted for, China's manufacturing sector has either narrowed the gap or overtaken the US. Of course America has not been sitting still. The continuing strong growth trend in US manufacturing suggests that reports of its industrial demise are greatly exaggerated.

~ ~ ~ ~

Industrial structure continues to change within each country as their internal technology and skills develop, but it also changes with increasing global trade. The most familiar shift is also the oldest, with industry shifting from farm to factory. The newest trend entirely upsets the pattern as industrial focus shifts from the factory to the office; extreme specialization breeds an ever growing and innovative services sector. And finally, a more hesitant kind of shift involves a back-and-forth between manufacturing and services. Figure 8.7 summarizes the manufacturing sector share within three modes of industrial shift.

DOI: 10.1057/9781137340313

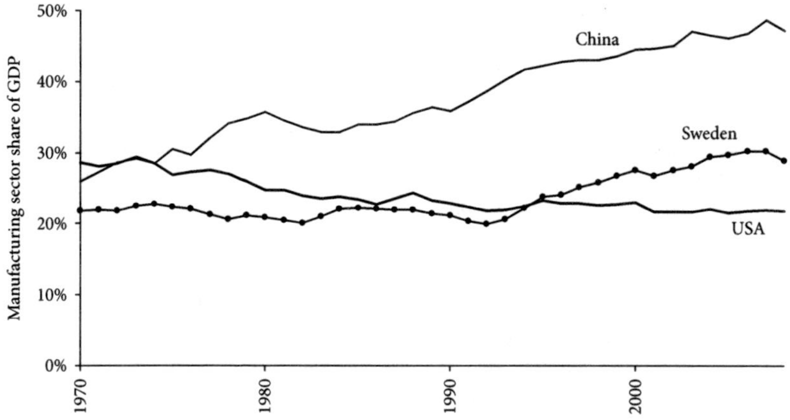

FIGURE 8.7 *Three country trends in manufacturing share, 1970–2008*
Source: Derived from UN National Accounts Database.

The US exemplifies the stable advanced economies around the world with its growing wealth ever less dependent on the factory. Sweden shows that changes can be reversed as its manufacturing share falls and rises in an ongoing exchange with the services sector. In a stark contrast, China has just now caught on to the example of the past American century, lifting itself out of poverty and into the middle phase of advancement by migrating to an urban manufacturing economy. Its manufacturing project is well underway, but it has barely started improving national wealth which still sits at 1/15th of the US level per person.

The US now signals an upcoming phase in the structure of modern economies. With the farm share stabilized at a low level, the US has completed the first part of its industrial shift. The current shift has run from manufacturing to services for a few decades, which raises the question of what the next phase might be. Will the services sector continue growing forever? Will manufacturing revert to a previous higher level?

Notes

1 This sample includes all countries in the EU KLEMS database with consistently structured TFP data. *Source*: derived from EU KLEMS Database Nov 2009 release. http://www.euklems.net/euk09i.shtml See 'Basic' Files. (Accessed February 19, 2011.)

DOI: 10.1057/9781137340313

2 The Western world includes Canada, United States, Australia, New Zealand, Western Europe, and Scandinavia. Their agricultural share fell only from 3% of GDP to 2% over the 38-year period. In contrast, GDP per capita increased by about 100%.

3 The correlation is −0.459 but improves even further with a two-year lag.

4 The correlation is +0.435.

5 Arnold, Bruce, *CBO Memorandum: Causes and Consequences of the Trade Deficit: An Overview*, The United States Congressional Budget Office, March, 2000.

6 Manufacturing output growth = 2.2% per year. Manufacturing employment growth = 0.48% per year. Population average growth = 0.93% per year.

7 In 2012 the mining-oil-gas sector employed 300,000 people out of a national total of 17.5 million. Derived from Statistics Canada CANSIM Database table 282–0008.

DOI: 10.1057/9781137340313

9
Where the Jobs Are

Abstract: *Industrial structure shifts along with labor productivity within an economy. But shifting geographic trade patterns also affect structure. In turn, both productivity and trade influence employment levels but not necessarily in the expected ways. Farm and factory output is rising in most countries. Canada's factory jobs rise and fall mainly with the dollar exchange rate. Strong German exports do not reduce unemployment. Productivity is reducing factory jobs in China.*

Some aspects of the US jobs crisis are disproved. Contrary to popular opinion, overall unemployment has improved as the trade deficit increased since the 1980s; service sector wages pay more and rise more than other sectors; production jobs and management jobs were equally affected by the recession.

Key words: China employment; job crisis; job trend; productivity; US unemployment

Atikian, Joe. *Industrial Shift: The Structure of the New World Economy*, New York: Palgrave Macmillan, 2013. DOI: 10.1057/9781137340313.

Exchange, exports, and deficits

Unemployment presents one of the most controversial aspects of national economies, as it affects people's daily lives as well as the political mood and social attitudes. Governments can manipulate unemployment figures to give a more favorable impression of their policies, for instance by changing the definition of the workforce or by issuing exaggerated positive reports followed by quiet negative corrections. These figures may attract suspicion or, worse yet, may be completely dismissed if they do not align with personal experience. In light of these grave difficulties, the following discussion focuses mostly on the *number of jobs* in order to avoid contentious unemployment data, and it addresses specific narrow issues instead of overarching ideas. Of course the downside of using jobs data is that it ignores population growth, but that is the inherent trade-off in avoiding the pitfalls of unemployment data. The idea is to illustrate a few concepts that connect job patterns to shifts in industrial structure, both domestically and globally. Are manufacturing jobs in decline? Are job numbers affected by imports and exports? Has the shift from manufacturing to services turned high-paying jobs into low-paying jobs? Is China gaining jobs at the expense of the West?

A brief look at manufacturing jobs in Canada will give an idea of why the prevailing view of industrial shifts can be so pessimistic. The repeated rise and fall in Canada's factory jobs makes it difficult to form a clear image of the general situation. The personal hardships created by episodes of major job losses understandably persist in the media and in memory, whereas new job celebrations often last for the day before fading away. Those who lose jobs in one multi-year cycle may not regain their jobs in the next cycle as some firms may be destroyed. Meanwhile new firms making new kinds of products may be created but remain out of reach for workers from vanishing industries, so the idea of general decline is very real to many workers.

Cycles in the number of jobs can last ten to fifteen years. As Figure 9.1 shows, the number of jobs lost and gained in a single cycle can approach half a million. But contrary to the more pessimistic view, the general trend in Canada's manufacturing jobs has been rising at least since 1976.

As expected, a stronger Canadian dollar reduces exports and employment.[1] But the chart also reveals a major complication that affects the

DOI: 10.1057/9781137340313

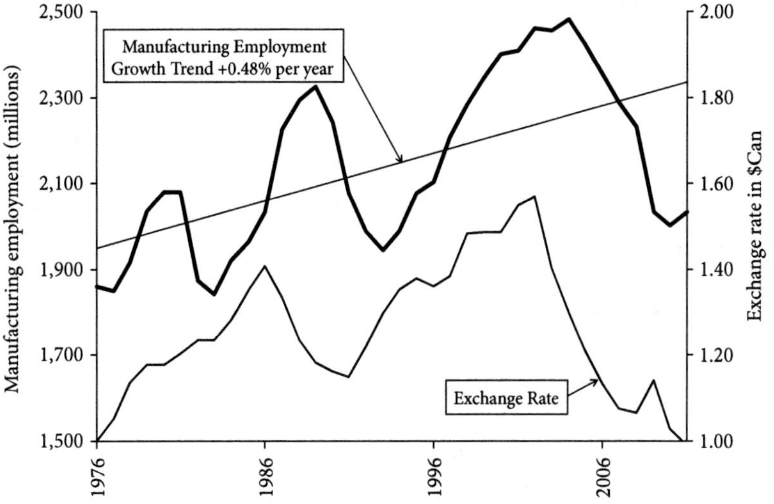

FIGURE 9.1 *Trend in Canadian manufacturing jobs and exchange rate, 1976–2011*
Source: Derived from Statistics Canada CANSIM Table 282–0008

way perceptions are formed: the uneven relation between the number of factory jobs and the US–Canada currency exchange rate. People have enough difficulty keeping track of the fluctuating number of jobs at each cycle, but what's more, a change in the currency affects employment only *after* two to three years, making the situation even more unclear for casual observers. Finally, even though the number of manufacturing jobs tends to increase, it increases at less than the population growth rate.

The result of these fluctuations is a great deal of anxiety about the death of manufacturing, as well as rancor about the causes. The blame for job losses often goes to poor government policy, corporate greed, union greed, and economic recession. Very little popular credence goes where it should, to the exchange rate, which can rise and fall 20% to 50% in a few years. That degree of change represents a major recurring shock to Canada's export prices, making the manufacturing sector similarly vulnerable to recurring employment shocks. Not every country experiences the same mechanism because each has its unique mix of trade, structure, and currency factors, but this just reinforces the point that there are no simple universal answers. Nor is there a universal trend of lower employment in manufacturing. See the box on Manufacturing employment by country below.

DOI: 10.1057/9781137340313

Focus on manufacturing employment by country

The World Bank and OECD post online employment figures for many of their member countries covering the past decade or more. Since 2003, the World Bank shows very mixed employment trends in manufacturing, with some countries and regions gaining jobs while others remain stable or shed jobs in the sector. A rising percentage of workers employed in manufacturing is seen within the developing countries of regional groupings such as the Middle East and North Africa as well as Europe and Central Asia. Some examples of individual countries with rising factory employment include Chile, Brazil, Mexico, India, Indonesia, and Poland.

The bank shows that the High Income Country group had a falling manufacturing employment trend since 2003; examples include South Korea, US, Japan, Sweden, and Germany.

Some countries including Thailand and Russia have had a decade with no change in the sector.

The OECD currently has 34 member countries and provides online data for 30 of them. Since 2000, 18 countries show declining trends, including the US, Denmark, Portugal, France, and Japan. Rising manufacturing employment was seen in New Zealand, Turkey, Mexico, Poland, Canada, and Australia. Stable employment in the sector was seen in Czech, Belgium, Switzerland, Iceland, Luxembourg, and Norway. No data are posted for Israel, Estonia, Slovenia, and Chile.

The 2008 global financial crisis also had mixed effects. The US and UK had a declining factory jobs trend for decades, and the crisis intensified the losses. Greece and Ireland had rising factory jobs, and the crisis severely reversed the trend. Australia and Turkey's rising trend went on largely unaffected.

Sources: World Bank http://data.worldbank.org/indicator/SL.IND.EMPL.MA.ZS
OECD http://stats.oecd.org/

~ ~ ~ ~

Germany has been touted as another global manufacturing and export powerhouse, recently noted for its trade surplus with China at a time when most of its competitor countries run a deficit (see Figure 9.2). This fact creates a lasting popular image of a strong economy worthy of emulating. This image, however, disregards Germany's overall unemployment

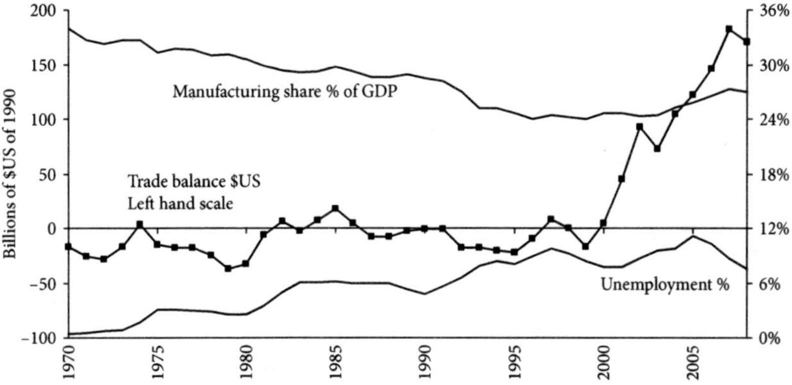

FIGURE 9.2 *Trend in German unemployment and trade surplus, 1970–2008*
Source: Derived from UN National Accounts Database and World Bank Indicators Database.

trend that has continually worsened to very high levels over the past four decades.

Germany's industrial structure gradually shifted toward services, as did that of Canada, Norway, the US, and others, with its manufacturing share falling by 7% of GDP over the period. Regardless of the falling manufacturing share though, real factory output grew by 80%. And while that output grew, exports exploded starting around 2000. Exports doubled from a stable long-term level of about one-quarter of GDP to nearly one-half. These rising exports would usually be expected to feed back into rising employment, but this did not happen. Employment continued its long gradual decline that started well before the 1990 German re-unification. This long-term deteriorating job trend seems to have been untouched by the export boom, again disproving the supposed positive connection between strong exports and jobs.

Stating this result may be interpreted as a claim that exports cause unemployment. This kind of interpretation is not intended because many other factors are known to be involved; claiming just one factor to be the cause is arbitrary. Rather, the justifiable conclusion is more limited in scope: strong exports do not assure low unemployment.

~ ~ ~ ~

America has a serious trade deficit with the rest of the world, and is said to be falling behind in manufacturing. Increasing imports of cheap goods

DOI: 10.1057/9781137340313

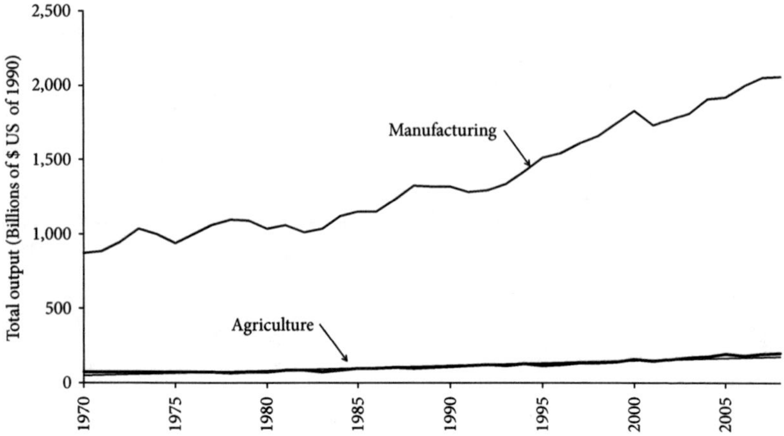

FIGURE 9.3 *US two-sector output trend, 1970–2008*
Source: Derived from UN National Accounts Database.

supposedly link to fewer goods being made domestically. This decreases output and employment, which in turn reduces tax revenues and raises government debt. Finally, lower tax revenues lead to government job cuts, forming a vicious cycle.

This mechanism sounds reasonable at first blush, but it should be tested by looking at manufacturing output and total unemployment:

- If the trade deficit is caused by imported goods displacing domestic goods, then the data should show that manufacturing output is falling.
- If imports are killing jobs, then the data should show unemployment to be rising.

Figure 9.3 shows US output in the two main industrial sectors, manufacturing and agriculture.

US farm output has nearly tripled since 1970 while factory output more than doubled in that period. The periods of slight decline are clear: the OPEC oil crises of 1973 and 1979; the early 1990s recession; 9–11–2001. The overall trend over the past several decades simply does not show a decline in manufacturing. Instead, the average annual growth rate for the sector was 2.7%. This growth rate was just under the GDP growth rate of 3%, which merely confirms that even though manufacturing output is growing, it makes up a slightly declining share of GDP.

DOI: 10.1057/9781137340313

The most unexpected result might be that factory output increased the most soon after trade with China opened up in 1980. The chart shows a stronger rise in the US manufacturing trend in 1982, a mere two years after China attained Most Favored Nation trade status.

What about employment? Aren't imports, especially from China, eliminating US jobs? Figure 8.1 shows the US trade balance with China. The shift is strong and clear: imports and exports were in balance until the early 1980s, and then the US fell into a large trade deficit. US exports to China grew rapidly, but imports from China grew even more.

This shift to a trade deficit happened very soon after China was granted normalized trade relations with the US in 1980, which fits with expectations. But notice that unemployment has a clear and unexpected trend. Between 1950 and 1980, trade with China was virtually non-existent, but the US unemployment trend was rising. Then the surprise: the trade deficit with China accelerates but unemployment falls. Now it seems very unclear that importing cheap Chinese goods could cause unemployment because, in fact, the data show unemployment went the opposite way.

None of this implies that the trade deficit or cheap imports have had no effect. It is true that they displace firms and entire industries, which in turn causes localized economic decline and job loss. But over the entire US economy, the expected link between imports and job losses has not materialized.

US jobs in crisis

Recent employment data reflect the effects of the financial crisis on job shifts within the US industrial base. Weakness in the US labor market persisted into the fourth year of the recession with the most intense shift hitting manufacturers. The gradual decline in the manufacturing share of GDP continues a decades-long trend that has accelerated since 2008. Out of 12.6 million production jobs in 1999, 4.4 million jobs were lost by 2010. There is no doubt about the severity of the impact on this sector, but another common feature of recessionary job losses is the idea that low-paying service sector jobs replace high-paying factory jobs. So-called burger flipping jobs are expected to replace unionized work or skilled work. But besides the recession, this echoes the job deterioration expected due to the rise of China and globalization. It seems reasonable

DOI: 10.1057/9781137340313

that importing manufactured goods would leave fewer jobs, and that the remaining jobs would be of lower value. So, the task here is to address one specific question: how many high-paying jobs were lost and replaced by low-paying jobs? The answer will help to clarify the kind of impact that comes with industrial shifts.

To find the shift between high-paying and low-paying jobs, the major job categories were tallied for 1999 and 2010 in Table 9.1. The table is ranked by hourly wage in 2010 (excludes self-employment).

How has the overall balance of high and low wage jobs changed? Out of approximately 127 million jobs in 1999, the overall change by 2010 was a small loss, −0.14 %. But the expected decline in high wage jobs did not materialize. The workforce lost nearly 3 million low wage jobs, and gained nearly 3 million high wage jobs. This turns out to be exactly the opposite of the widely expected decline into low-paying jobs as the country's industries shift toward services.

Wages earned in production jobs (manufacturing) ranked nearly centered between the high and low halves of the workforce. The surprise here was that the production category is not in the high-paying half of the workforce. Instead, it is ranked second in the lower wage segment, just below the "office and administrative support" category. It seems that the high pay of production jobs is at least partly a myth, while the popular image of low-paying food service jobs is confirmed by its last-ranked position in the list. The 11 million food service jobs are indeed low-paying service sector jobs. But as a counterbalance, the 11 million service sector jobs in engineering, computers, and business operations are high paying, at over triple the hourly wage found in food service.[2]

Production jobs showed the largest absolute loss of any category at nearly 4.4 million. More surprisingly, the next largest change in job numbers was in the highest paying category, management, with a loss of just over 2 million. This includes a 54% loss of CEO positions, and a 66% loss of advertising management jobs. As a proportion of total jobs in each category, production workers and managers lost the same amount, 25% since 1999.

The biggest job gaining areas in the high-paying category were business & financial operations, and health care practitioners. Most service sector job gains came in health care, education, computer, and business operations, all of which pay much higher wages than paying food service jobs.

After finding that the number of low-paying jobs decreased and high-paying jobs increased, further questions about income will naturally arise.

DOI: 10.1057/9781137340313

TABLE 9.1 *Shifts in US jobs by sub-sector, ranked by 2010 wage, 1999 to 2010*

Occupation	Employment 2010	Hourly median wage, 2010 (USD)	Employment 1999	Hourly median wage, 1999 (USD)	Employment gain/loss	Real annual wage increase (CPI = 2.8%)
Food Preparation & Serving	11,027,340	9.02	9,687,970	5.96	1,339,370	3.5%
Farming Fishing & Forestry	408,040	9.44	463,360	6.20	−55,320	3.6%
Personal Care & Service	3,425,220	9.92	2,556,920	6.37	868,300	3.8%
Building Cleaning & Maintenance	4,175,550	10.81	4,274,200	6.58	−98,650	4.2%
Sales & Related	13,437,980	11.72	12,938,130	6.68	499,850	4.8%
Healthcare Support	3,962,930	11.90	2,970,780	7.36	992,150	4.1%
Transportation	8,547,980	13.66	9,538,820	7.50	−990,840	5.1%
Production	8,236,340	14.58	12,620,920	8.21	−4,384,580	4.9%
Office & Admin Support	21,503,800	14.77	22,562,480	8.69	−1,058,680	4.5%
Protective Service	3,187,810	17.63	2,958,730	8.30	229,080	6.5%
Arts, Entertainment & Media	1,716,640	20.61	1,551,600	9.87	165,040	6.3%
Community & Social Service	1,901,180	18.89	1,404,540	10.56	496,640	5.0%
Construction & Extraction	5,072,530	18.79	5,938,860	10.68	−866,330	4.8%
Education Training & Library	8,457,870	21.97	7,344,830	10.69	1,113,040	6.2%
Installation, Maintenance	4,928,960	19.29	5,140,210	10.88	−211,250	4.9%
Healthcare Practitioners	7,346,580	28.12	6,001,950	13.71	1,344,630	6.2%
Life Physical & Social Science	1,064,510	28.14	909,530	14.51	154,980	5.7%
Legal	992,650	35.86	858,320	16.38	134,330	6.7%
Architecture & Engineering	2,305,530	33.95	2,506,380	17.53	−200,850	5.7%
Computer & Mathematical	3,283,950	35.44	2,620,080	18.61	663,870	5.5%
Business & Financial Operations	6,090,910	29.17	4,361,980	20.08	1,728,930	3.2%
Management	6,022,860	43.96	8,063,410	27.68	−2,040,550	3.9%
Total	127,097,160		127,274,000		−176,840	

Source: Derived from US Bureau of Labor Statistics, Occupational Employment and Wage Estimates.

Hasn't the industrial shift toward services reduced people's income? Many reports repeat the fact that median household income has stagnated or fallen in recent decades, for example quoting from the US Census Bureau that "Real median household income fell between 2007 and 2008, and the decline was widespread."[3] ('Real' refers to inflation adjusted). A common criticism of this conclusion is that the number of earners in a household falls as the divorce rate rises, meaning that the decline in earnings is exaggerated: instead of one household earning $80,000 a divorce results in two households with an average of $40,000 each. Even with the same total income in the economy, divorces produce a lower median income. The aforementioned report recognizes this problem but does not address it,[4] so a brief return to Table 9.1 is in order.

Instead of median household income, the table shows median hourly wage rates for each job category. Over the 12-year period, wages in each category increased although inflation did likewise. Adjusting for inflation gives the result that real wages increased in each category for each year, again, not the outcome expected according to the commonly quoted household income concept from the Census Bureau. Inflation diminished the value of wages by 2.8% per year but wages in every category more than made up for that, rising between 3.5% and 6.7% per year depending on the job category.[5] And again, the finding here is that the industrial shift from manufacturing to services has been beneficial to workers' total wage income.

East meets West

Even some quirky cultural references demonstrate people's concern about the global industrial shift, and the US–China shift in particular. One commentator recently lamented that a popular activity among American youth is to create video clips of themselves car-surfing or exhaling clouds of cinnamon powder, and then he warned that China will eat our lunch. In other words, if the West does end up in decline, it will bear its own responsibility for giving up on ambition, education, and discipline.

Daily warnings of a more serious kind continue about off-shoring, outsourcing, and job losses in the advanced Western economies, but they ignore a clear potential benefit of China rising. Just as America raised its own industrial capability to overtake Britain as the world's largest economy in 1906, it transformed itself into a market for global imports.

DOI: 10.1057/9781137340313

Imports were not a major share of the US economy, but the economy was so large that it nevertheless became an important global market. Likewise, as China becomes a more sophisticated industrial economy, it will present itself as an important market to the rest of the world. Indeed, private firms that are accused of outsourcing to China merely for its cheap labor have long since anticipated the Chinese market. But Western firms opening branches in China knowingly face a dilemma. On the one hand, such firms are usually required to join a Chinese partner that holds a majority or controlling interest in the local operation. They are also required to divulge the technological information that underpins their product. On the other hand, these firms look forward to the prospect of an enormous new market for their products, a market that would otherwise be impossible to enter.

The choices are clear. Either firms can keep their technology to themselves and be shut out of China, or they can divulge it for the opportunity to grow within China. If the choices are clear, then so too are the risks. Although US median wages and the number of service jobs have increased, nothing guarantees that they will continue increasing. The sheer scale of China's industrial capacity may overwhelm America's production regardless of future increase or decrease in outsourcing. Services exports from China may be the next wave. But one certainty is that relaxing the push for innovation or closely guarding secrets will not help to create jobs in the global market.

The American auto market provides a specific analogy. The world's most popular and coveted luxury car brands are imported from Germany, a country with the unimaginable disadvantage of having been virtually destroyed in the Second World War. It had been 'de-industrialized' by military means twice in the past century, and still managed to rebuild its own auto industry to become the most respected in the world. Facing the difficulty of an open market against all comers in the US, Daimler-Benz, BMW, and Audi established an export market in this much larger overseas economy, a former military enemy that speaks a completely different language. In addition, the luxury auto segment attracted special taxes that might have discouraged its consumers.

Even against all of these barriers though, the top German brands built such a strong following that they eventually tilted toward manufacturing some of their cars in the US (being near customers is beneficial, and local content protects against currency shifts). And the same brands are now coveted in China, which has become their newest mass export market as

DOI: 10.1057/9781137340313

well as a new manufacturing center. The US shares most of these difficulties as it faces China today: language, scale, distance, competitors. It may overcome all of these hurdles, but opportunities in China will likely be lost without major innovations in technology and trade regimes.

Of course, the above analogy is imperfect. For post-war Germany, the US already existed as an export market, and shifted to a branch manufacturing location. Intellectual property rights were protected by law. In contrast, China presents the opposite situation: a manufacturing economy that seems headed toward becoming an export market with less protection. The sequence of events is vastly different in the two cases, but whether or not America follows Germany's path to success in China will still rest on the state of US technology. Its success will not stem from trade restrictions or complaints against the Chinese government's currency manipulation. The geographic implications for employment may not be entirely clear at this early stage, but part of the jobs outcome may depend on the choice between video clips and creating the next technological advances to sell within the world's largest imminent market.

Productivity strikes back

Productivity can easily be mistaken for working harder, meaning greater effort, higher speed, or longer days. While each of these can increase production output, they also have limits beyond which waste increases and more errors are made. They also have physical limits, for example, in that nobody can work five times faster than their usual pace. Instead, significant productivity improvements come from increased knowledge, specialization of tasks, and an increasing use of equipment. One of the oldest accounts of productivity improvement comes from Adam Smith in 1776.[6] He observed that one man could make no more than twenty pins in a day on his own, but by assigning one man to perform each step needed to make a pin, each man could in effect make 4,800 pins. In that example, productivity had risen by 240 times, almost infinitely beyond the few percentage points more that one man could make by "working harder". But when productivity rises, two options crop up: make more pins, or employ fewer people to make the same amount. People naturally worry about the number of jobs being reduced by productivity, and they do so to the point of caricature, at which time robots make everything and everyone is unemployed.

DOI: 10.1057/9781137340313

Some economists argue that raising productivity is the only way in which the standard of living can improve. Many people will argue against that view because they see it leading to higher unemployment. Both sides have merit but instead of a detailed discussion, a simple example gives an overview of the true result. When telephones were first introduced, a person manually plugging the desired wires together connected each call. These Switchboard Operators were displaced by the new technology in today's telephone systems, the automatic switches that connect every phone call or text message. A popular story in the 1980s was that if the manual operator method were still used to connect all the calls made today, it would take more than the entire workforce just to perform that single task. So of course, new technology has eliminated every existing switchboard job, but workers moved on to other new fields without a drastic change in the number of jobs.

The opposite point of view also illustrates the connection between jobs and productivity. Wouldn't the prohibition of technological change be good for employment? Government policy could totally eliminate unemployment with one simple rule: no machinery in coal mining or corn harvesting. That means digging coal and harvesting corn by hand. It can be done, and it was done this way for years. It is still done in some places. It simply means undoing technological progress in two small industry sectors, but the effect on jobs would be astonishing. The entire jobless workforce could be working tomorrow, and yet nobody would want to do it. People certainly favor technical progress such as mechanized corn harvesting when it makes work easier, less boring, and safer, but prohibiting change makes that impossible.

Productivity generally reduces employment in some industries, so more workers are available for new industries. This is the root of the industrial shift. For example, US manufacturing output grew at over 2% per year from 1970 to 2008 while the population grew at only 1% per year; that extra output came from increasing productivity and not from more workers. Most of a new generation would work in offices instead of factories. Meanwhile, China's manufacturing output increased rapidly in the 1980s, much of it being exported to the US. But contrary to the expected job loss from these two trends, the overall US unemployment trend declined for the next 25 years. This happened at least partly because lower prices on Chinese made consumer goods allowed consumers to spend on higher value domestic services and manufactured goods such as cars.

DOI: 10.1057/9781137340313

If this still does not seem plausible, consider that the US has seen ongoing technological progress in manufacturing for well over 100 years. So if job losses were higher than job gains, unemployment would continually rise. Instead, with ever rising productivity, unemployment merely fluctuates between about 3% and 7%. So far, rising productivity and imports have shifted US jobs between sectors without decimating the workforce or its median wages. But ultimately, the industrial shift in the US is part of a larger shift in the world economy, one that has been ongoing forever.

Just as the US economy overtook the UK in 1906, China appears set to assume the top rank in economic output today. And just as America grew on its deep trade ties with Britain, China is growing on its deep trade ties with America. One important difference has to do with starting points. America attained its economic dominance as a new country on a lightly populated continent. It absorbed and employed a growing new workforce, an enormous change in where the jobs were in the world. In contrast, China's approaching dominance is in fact an economic re-birth or re-awakening of a once great and populous economy. In the past 40 years, China added twice as many to its population as the US added over the course of a century.[7] In this light, it seems conceivable that China's industries could utterly crush America's simply by replacing the US workforce in the global marketplace. So, are China's factory workers in fact replacing factory workers in America? Will this be the next great change in where the jobs are?

China's factory employment increased by several millions from 1990 to 1995 but declined for the next five years. See Figure 9.4. America's factory job numbers were virtually unchanged through these ten years, possibly casting doubt on the idea of losing US jobs to China.

But before urging any conclusion, a caution about the numbers from China. The two peaks in China's factory employment in 1997 and 2007 do not indicate a real trend, but a change in the way the government calculates the figures. A great deal of uncertainty surrounds these official figures, but an investigation by the US Bureau of Labor Statistics concluded that China lost 8.5 million factory jobs since 1995.[8] It seems that the general factory employment trend for both countries has been in gradual decline.

Can the loss of US manufacturing jobs in this period be attributed solely to China? No. First, China's falling job trend plainly runs counter to the idea of displacing US jobs. Second, if China really was the sole cause, then there could have been no effect whatsoever from rising US productivity, or from jobs shifting to the services sector. Since this is not true,

DOI: 10.1057/9781137340313

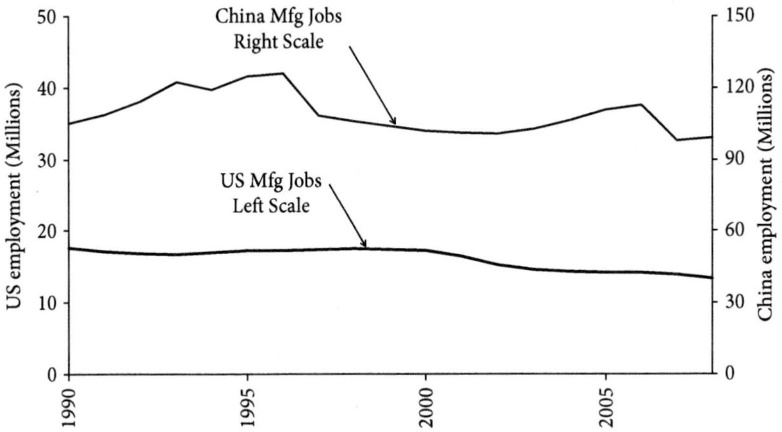

FIGURE 9.4 *Trend in US and China manufacturing jobs, 1990–2008*
Source: Derived from US Bureau of Labor Statistics and BLS Monthly Labor Review.

China cannot be the sole cause. Nevertheless, this does not prove that the US lost no jobs to China. Instead it demonstrates that China has shed manufacturing jobs due to higher productivity, and is not immune to the same forces of change that the US sees. It also matches Germany's experience, demonstrating that China's high and rising exports do not prevent its own employment shifts. The implication is that the US should be wary of pursuing exports as the key means to addressing unemployment.

So US factory jobs were more likely lost to a combination of three causes. First, US manufacturing productivity has continually risen, and the sector has simply reduced its demand for labor. China's recent experience was likely the same in this regard. Second, some US factory jobs shifted to US services in a decades-long trend. There were about 2.3 million new service sector jobs added since 1999, and some were filled by manufacturing workers. Third, some US jobs certainly shifted to China; some US firms closed domestic plants and opened in China specifically to reduce labor cost. It turns out that the jobs are everywhere.

The most obvious remaining question must be that if China's labor cost is so low, why do all major manufacturing firms simply not move there and increase their profits? After all, the average Chinese factory wage stands at about 4% of the average American wage,[9] meaning that labor is virtually free. This question has been asked for decades and the answers have grown correspondingly complex. But the most basic part of the answer can offer a clue about the kind of gap that profit seeking firms are facing. If China's wages are low, then its productivity is also likely to be low, and firms must obviously pay for more hours of labor to produce the required amount of goods. So, how does China's productivity compare? Return to the first topic in this discussion, agriculture. The average US farm worker produces about $51,000 worth of goods per year whereas the average farm worker in China produces about $545 worth per year.[10] At this rate, a firm would need about 93 times the number of workers to generate the same output, which eliminates the low wage advantage. To compensate for the large number of workers, new machinery would be introduced which in turn requires capital investment along with ongoing training that eventually leads to higher wages. Operating in the lowest wage country may not always be the most profitable or effective. The West certainly has not fallen behind by this measure.

Notes

1 The correlation with a three-year lag is very strong at 0.85.

2 The dividing line between high and low wage categories was $15 per hour. The idea is to find the wage below which one-half of the jobs fall. This is somewhat arbitrary as there is no dividing line that produces an exact 50–50 split in the number of jobs. The line can be moved at least one level higher or lower than $15 without significantly affecting the outcome.

If the midpoint of the wage range (i.e., $26.49) is used as the high-low cut-off, the result is still that 2 million low wage jobs were lost and 1.8 million high wage jobs were gained. It is difficult to avoid the conclusion that wages have risen with the shift toward services.

3 C. DeNavas-Walt, et al., U.S. Census Bureau, Current Population Reports, P60–236, *Income, Poverty, and Health Insurance Coverage in the United States: 2008*, U.S. Government Printing Office, Washington, DC, 2009. pg 1. http://www.census.gov/prod/2009pubs/p60–236.pdf

4 C. DeNavas-Walt, et al., U.S. Census Bureau, Current Population Reports, P60–236, *Income, Poverty, and Health Insurance Coverage in the United States: 2008*, U.S. Government Printing Office, Washington, DC, 2009. http://www.census.gov/prod/2009pubs/p60–236.pdf., p. 8; see note 13. "The distribution of household income is influenced by many factors, such as the number of earners and household size."

5 C. DeNavas-Walt, et al., U.S. Census Bureau, Current Population Reports, P60–236, *Income, Poverty, and Health Insurance Coverage in the United States: 2008*, U.S. Government Printing Office, Washington, DC, 2009. http://www.

DOI: 10.1057/9781137340313

census.gov/prod/2009pubs/p60-236.pdf., p. 28. Derived from the US Census Bureau research series of the consumer price index.

6 A. Smith, *An Inquiry Into the Nature and Causes of the Wealth of Nations*, Prometheus Books, Amherst NY, 1991, p. 10.

7 From 1970 to 2010 China added 500 million people. From 1910 to 2010 the United States added 216 million.

8 "No matter how the official data are adjusted, China's total manufacturing employment dropped by around 8.5 million or more from 1995 to 2000." J. Bannister, *Manufacturing Employment and Compensation in China*, US Bureau of Labor Statistics, November 2005, p. 8. http://www.bls.gov/fls/chinareport.pdf

9 J. Bannister, G. Cook, "China's Employment and Compensation Costs in Manufacturing through 2008," US Bureau of Labor Statistics, *Monthly Labor Review*, March 2011, p. 39.

10 http://data.worldbank.org/indicator/EA.PRD.AGRI.KD/countries (Accessed January 28, 2013).

DOI: 10.1057/9781137340313

Appendix A: UN Least Developed Countries

Afghanistan
Angola
Bangladesh
Benin
Bhutan
Burkina Faso
Burundi
Cambodia
Central African Republic
Chad
Comoros
Democratic Republic of the Congo
Djibouti
Equatorial Guinea
Eritrea
Ethiopia
Gambia
Guinea
Guinea-Bissau
Haiti
Kiribati
Lao People's Democratic Republic
Lesotho
Liberia
Madagascar
Malawi
Mali

DOI: 10.1057/9781137340313

Mauritania
Mozambique
Myanmar
Nepal
Niger
Rwanda
Samoa
São Tomé and Príncipe
Senegal
Sierra Leone
Solomon Islands
Somalia
South Sudan
Sudan
Timor-Leste
Togo
Tuvalu
Uganda
United Republic of Tanzania
Vanuatu
Yemen
Zambia
(see http://www.unohrlls.org/en/ldc/)

DOI: 10.1057/9781137340313

Appendix B: Compounded average annual growth of Total Factor Productivity, selected OECD countries

Country	Agriculture TFP growth (%)	Manufacturing TFP growth (%)	Period
Australia	3.0	0.8	1982–07
Austria	3.1	3.1	1980–07
Belgium	2.1	1.3	1980–06
Czech	1.4	3.3	1995–07
Denmark	2.5	0.2	1980–07
Finland	3.3	3.2	1970–07
France	3.5	1.6	1980–07
Germany	4.8	2.0	1991–07
Hungary	4.7	3.5	1995–07
Ireland	2.2	2.4	1988–07
Italy	3.7	1.5	1970–07
Japan	0.4	2.2	1973–06
Netherlands	2.2	1.9	1979–07
Spain	3.1	0.6	1980–07
Sweden	2.8	4.0	1993–07
UK	2.2	1.8	1970–07
US	3.1	2.2	1977–07

(*Note*: Productivity in agriculture grows faster than manufacturing in most cases.)
Source: Derived from EU KLEMS Database Nov 2009 release.

DOI: 10.1057/9781137340313

Appendix C: Rank of GDP per capita in 2008, all countries of the world (in constant US$ of 1990)

183,804	Monaco
83,194	Liechtenstein
56,975	Luxembourg
55,786	Bermuda
42,683	Norway
41,176	British Virgin Islands
40,695	Switzerland
40,012	Finland
39,286	Sweden
39,091	San Marino
35,545	Denmark
35,434	Iceland
30,682	Ireland
30,363	United States
30,055	Japan
29,890	Austria
28,415	Netherlands
28,246	Germany
28,132	Canada
28,036	Cayman Islands
27,433	Belgium
27,006	France

Continued

26,723	Australia
26,046	Andorra
25,634	Qatar
25,496	United Kingdom
24,525	Greenland
24,516	Singapore
23,855	Italy
23,029	Hong Kong
21,360	United Arab Emirates
19,741	Spain
18,484	Israel
17,720	Turks and Caicos Islands
17,575	Kuwait
17,506	New Zealand
15,083	Cyprus
14,657	Greece
14,504	South Korea
14,202	Bahrain
13,808	Aruba
13,322	Anguilla
13,031	Bahamas
12,980	Puerto Rico
12,953	French Polynesia
12,837	Brunei Darussalam
12,826	Netherlands Antilles
12,616	New Caledonia
11,799	Malta
10,614	Trinidad and Tobago
10,105	Portugal
9,922	Libya
9,693	Antigua and Barbuda
9,403	Seychelles
8,854	Barbados
8,094	Saudi Arabia
7,473	Argentina

Continued

7,365	Oman
6,605	Equatorial Guinea
6,145	Saint Kitts and Nevis
5,538	Turkey
5,165	Hungary
5,147	Chile
5,129	Gabon
5,018	Cook Islands
4,984	Malaysia
4,932	Botswana
4,892	Montserrat
4,781	Panama
4,641	Uruguay
4,594	Palau
4,562	Mauritius
4,487	Mexico
4,228	Brazil
3,964	Costa Rica
3,794	Saint Lucia
3,725	South Africa
3,716	Grenada
3,582	Bulgaria
3,550	Saint Vincent and the Grenadines
3,547	Cuba
3,399	Poland
3,380	Dominica
3,191	Belize
2,926	Venezuela
2,917	Algeria
2,907	Thailand
2,848	Tunisia
2,814	Namibia
2,785	Iran
2,549	Dominican Republic
2,498	Romania

Continued

DOI: 10.1057/9781137340313

2,384	Peru
2,333	Maldives
2,273	Jamaica
2,211	Colombia
2,174	Fiji
1,976	Suriname
1,954	Cape Verde
1,836	Tonga
1,803	China
1,794	Jordan
1,776	Marshall Islands
1,749	Lebanon
1,736	Angola
1,717	Morocco
1,705	Tuvalu
1,559	Sudan
1,535	El Salvador
1,509	Ecuador
1,487	Swaziland
1,423	Sao Tome and Principe
1,390	Syria
1,371	Micronesia
1,352	Bhutan
1,298	Albania
1,253	Indonesia
1,215	Congo
1,167	Paraguay
1,144	Egypt
1,138	Vanuatu
1,081	Samoa
1,019	Papua New Guinea
1,016	Honduras
1,014	Occupied Palestinian Territory
1,012	Sri Lanka
1,009	Guatemala

Continued

DOI: 10.1057/9781137340313

984	Bolivia
978	Mongolia
975	Philippines
972	Cameroon
951	Guyana
949	Senegal
948	Nicaragua
893	Nauru
847	India
843	Iraq
812	Djibouti
776	Gambia
775	Côte d'Ivoire
737	Nigeria
727	Pakistan
719	Mauritania
651	Solomon Islands
627	Ghana
556	Lesotho
546	North Korea
534	Guinea
527	Burkina Faso
490	Comoros
478	Zambia
476	Kenya
466	Zimbabwe
466	Mozambique
463	Rwanda
462	Benin
461	Mali
438	Laos
438	Bangladesh
437	Uganda
410	Central African Republic
400	Chad

Continued

DOI: 10.1057/9781137340313

377	Cambodia
344	Togo
337	Tanzania: Mainland
337	Niger
333	Afghanistan
311	Malawi
308	Nepal
276	Viet Nam
251	Madagascar
248	Haiti
179	Guinea-Bissau
178	Sierra Leone
160	Liberia
152	Burundi
118	Democratic Republic of the Congo
106	Somalia

Source: Derived from UN System of National Accounts.

DOI: 10.1057/9781137340313

Bibliography

Arnold, B., *CBO Memorandum: Causes and Consequences of the Trade Deficit: An Overview,* The United States Congressional Budget Office, March, 2000.

Bannister, J., *Manufacturing Employment and Compensation in China*, US Bureau of Labor Statistics, November 2005.

Campbell, D., *History, Culture, and Trade; A Dynamic Gravity Approach*, UC Davis Working Paper, 2010.

Food and Agricultural Organization of the United Nations, *Summary of Food and Agricultural Statistics 2003*, Rome 2003.

De Gregorio, J., *Economic Growth in Latin America: From the Disappointment of the Twentieth Century to the Challenges of the Twenty-First*, Central Bank of Chile Working Papers, No 377 (abstract), November 2006.

DeNavas-Walt, C., Proctor, B., and Smith, J., U.S. Census Bureau, Current Population Reports, P60–236, *Income, Poverty, and Health Insurance Coverage in the United States: 2008*, U.S. Government Printing Office, Washington, DC, 2009.

Dinopoulos, E., *North-South Trade and Economic Growth*, University of Florida, June 5, 2006.

FRBSF Economic Newsletter 2011–25, Federal Reserve Bank of San Francisco (FRBSF), August 8, 2011.

Mankiw, N. G., *China's Trade and U. S. Manufacturing Jobs*, Chairman of the Council of Economic Advisers Testimony Before the House Committee on Ways and Means, Washington, DC, October 30, 2003.

DOI: 10.1057/9781137340313

Kuznets, S., "National Income & Industrial Structure," *Econometrica*, Vol. 17, July 1949, The Econometric Society.

Kuznets, S., "Quantitative Aspects of the Economic Growth of Nations," *Economic Development and Cultural Change*, Vol. 5, No. 4, July 1957, University of Chicago Press.

Smith, A., *An Inquiry Into the Nature and Causes of the Wealth of Nations*, Prometheus Books, Amherst, NY, 1991.

Wortham, A., *Schumpeter and His Critics on the Future of Capitalism*, Working Papers in Economics E-88–27, The Hoover Institution Stanford University, July 1988.

DOI: 10.1057/9781137340313

Index

DOI: 10.1057/9781137340313

DOI: 10.1057/9781137340313

DOI: 10.1057/9781137340313

CPSIA information can be obtained at www.ICGtesting.com
Printed in the USA
LVOW08*1635310114

371834LV00006B/169/P

9 781137 342263